Translating Systems Thinking into Practice

A Guide to Developing Incident Reporting Systems

Translating Systems Thinking into Practice

A Guide to Developing Incident Reporting Systems

Natassia Goode
Paul M. Salmon
Michael G. Lenné
Caroline F. Finch

CRC Press
Taylor & Francis Group
Boca Raton London New York

CRC Press is an imprint of the
Taylor & Francis Group, an **Informa** business

First published in paperback 2024

First published 2019
by CRC Press
4 Park Square, Milton Park, Abingdon, Oxon, OX14 4RN

and by CRC Press
2385 NW Executive Center Drive, Suite 320, Boca Raton FL 33431

© 2019, 2024 Taylor & Francis Group, LLC

CRC Press is an imprint of Informa UK Limited

Publisher's Note
The publisher has gone to great lengths to ensure the quality of this reprint but points out that some imperfections in the original copies may be apparent.

Library of Congress Cataloging-in-Publication Data

Names: Goode, Natassia, author. | Salmon, Paul M., author. | Lenne, Michael G., author. | Finch, Caroline F., author.
Title: Translating systems thinking into practice : a guide to developing incident reporting systems / Natassia Goode, Paul M. Salmon, Michael G. Lenne, and Caroline F. Finch.
Description: First edition. | Boca Raton, FL : CRC Press/Taylor & Francis Group, 2018. | Includes bibliographical references and index.
Identifiers: LCCN 2018027158| ISBN 9781472436917 (hardback : acid-free paper) | ISBN 9781315569956 (ebook)
Subjects: LCSH: Industrial safety--Management. | Industrial accidents--Reporting.
Classification: LCC T55 .G596 2018 | DDC 658.4/7--dc23
LC record available at https://lccn.loc.gov/2018027158

ISBN: 978-1-4724-3691-7 (hbk)
ISBN: 978-1-03-292275-1 (pbk)
ISBN: 978-1-315-56995-6 (ebk)

DOI: 10.1201/9781315569956

Visit the Taylor & Francis Web site at
http://www.taylorandfrancis.com

and the CRC Press Web site at
http://www.crcpress.com

For Miles and Tristan

Contents

Preface

Within safety science, it is widely accepted that a systems thinking approach is required to understand, and ultimately prevent, incidents in contemporary workplaces. The basic premise of this approach is that safety in organizations is influenced by the decisions and actions of people at all levels of the work system (e.g. supervisors, managers, chief executives, safety officers, regulators, and government), not just workers directly involved in the incident. Incidents are therefore a product of multiple, interacting, factors across a work system (Rasmussen 1997, Leveson 2004).

This philosophy is by no means new, and several methods* have been developed to understand and analyze incidents from this perspective. Studies have demonstrated the advantages of these methods, over other more reductionist methods, in many domains. The core conclusion from this large body of work is that human errors, procedural violations, and technology failures typically result from poorly designed and managed work systems. Incident prevention strategies, or countermeasures, therefore need to focus on changing the conditions of work, rather than just reiterating the importance of compliance with existing procedures, or introducing new equipment, procedures, or training.

This approach is firmly embedded in incident reporting and investigation practices in commercial and military aviation. This continual cycle of feedback and systems improvement over many years has resulted in exceptionally safe operations in these settings.

In many other industries, there remains a significant gap between state-of-the-art research on incident analysis and prevention, and practice. This is particularly so in the context of incident reporting systems. Most incident reporting systems are designed on an ad hoc basis, without any reference to the research on accident causation models or analysis methods. Consequently, most incident reporting systems are only capable of capturing data about the injured person or the immediate context of the event, rather than the broader contributory factors across the work system. As a result, little is typically learnt from reports of incidents, and in many organizations, the costs of administering the incident reporting system outweigh any real safety benefits.

This book attempts to address this gap by describing a program of research in which we designed and implemented an incident reporting system that was (1) directly underpinned by the systems thinking approach; and (2) uses

---

* An accident analysis method is a systematic or established procedure for analyzing the data collected about an incident. Many methods also include a way of graphically representing the findings from the analysis in a diagram.

systems thinking analysis methods. In doing so, the book provides guidance on designing a practical, usable incident reporting system underpinned by a systems thinking approach. This guidance is based on state-of-the-art systems thinking theories and methods, the literature on good practice for incident and injury data collection, and the lessons learnt from developing, testing, and implementing an incident reporting system for the outdoor sector in Australia. The resulting system became known as the Understanding and Preventing Led Outdoor Accidents Data System (UPLOADS, at www .uploadsproject.org).

UPLOADS has been highly successful during its first three years of operation. Thirty-five organizations have contributed data to a national incident dataset, which is regularly analyzed by the research team and reported back to the sector. This demonstrates the strength of the approach that we have used to develop UPLOADS, and the sector's commitment to preventing incidents and injuries during outdoor programs.

The commitment and leadership of our industry partners has been central to the success of this work (see Acknowledgements). In the following section, one of our industry partners, Clare Dallat, who has more than 15 years of experience in risk management, describes how the need for a new approach to incident reporting was identified by the outdoor sector, and the factors motivating the ongoing commitment to the UPLOADS project.

Why Did We Need a New Approach to Incident Reporting?

Every year in Australia, thousands of LOA programs are conducted that all share one common goal – the provision of an enriching, outdoor experience where the risk of harm to our participants is minimized.

Over 2008 and 2009, a small group of Australian LOA practitioners met to discuss the injuries and fatal incidents that had occurred in our respective organizations. We realized we were experiencing similar events, and that the strategies adopted to prevent incidents were not always successful. We also realized that most incident reports, and indeed the resulting strategies, focused solely on the behavior of the staff and participants involved in the activity (e.g. retraining, reprisals, review of procedures).

After many conversations, we realized we needed a more in-depth understanding of the types, frequency, and magnitude of incidents experienced during LOA programs, and the factors contributing to them. This led to the sector committing to a new program of research, involving what is now a highly successful collaboration with researchers from the University of the Sunshine Coast, Federation University of Australia, Monash University, and Edith Cowan University.

The intention at the time (2009) was to review what was known about the contributory factors involved in LOA incidents. This initial piece of work revealed that little was known, in part, because there were few incident reporting systems. Critically, the incident reporting systems that did exist were unable to provide us with an in-depth understanding of what was causing LOA incidents. The outcome was a proposal to the Australian Research Council to develop and implement a new sector-wide incident reporting system that was aligned with the systems thinking approach.

The ensuing program of research has completely transformed the LOA sector's understanding of incident causation. UPLOADS has helped us, as leaders within the outdoor sector, grasp that multiple factors, actors, and decisions and interactions across the LOA system contribute to minor injury incidents and fatal events. Another major benefit from UPLOADS has been the real and factual insight into the types of activities associated with the most incidents. It was surprising to us that it is not the high risk, high adventure, high equipment intensive activities that experience the greatest number of injuries. Rather, it is less overtly risky activities such as free time, cooking, and bushwalking that have the most frequent incidents and injuries. With its rich descriptions of incident causation, UPLOADS has enabled the sector to consider, and implement, incident prevention strategies based on real data. This is a huge advancement for us, and will ensure that we are able to provide safe LOAs for generations to come.

As a safety practitioner, I would highly recommend this book to anyone interested in developing a practical and theoretically sound approach to incident reporting. It is my personal view that this program of research, and its practical translation has, and continues to prevent, real harm, whilst also enabling participants to continue to achieve the significant and meaningful outcomes associated with participation in LOA programs across Australia.

Who Should Read This Book?

This book is intended to provide a guide for practitioners and researchers wishing to develop new incident reporting systems, as well as those wishing to evaluate and improve their existing systems. Students studying safety management will also find the book useful.

The book encompasses the development of both internal incident reporting systems established within a specific organization, and sector- or industry-wide systems established to collect and analyze incident reports from multiple organizations. The processes required to develop and evaluate both types of systems are largely the same; the additional factors to consider in the design of sector-wide systems are highlighted throughout the book.

Finally, the reader interested in general principles, methods, and philosophies for safety management will find sections of the book interesting. These include overviews of systems thinking, accident analysis methods, and testing the reliability and validity of accident analysis methods.

Why Should You Read This Book?

Designing, testing, and implementing an incident reporting system is a massive undertaking. Many organizations are concerned that their existing incident reporting systems do not enable them to understand and prevent incidents, even after much time and effort has been spent on developing the incident reporting software and training staff to use it. The guidance in this book is intended to help readers to work more collaboratively with software developers to get a better outcome.

This book breaks incident reporting systems down into their most basic elements, and describes how to design these elements to ensure the resulting system is practical for users, produces good quality data and analyses, and reflects the principles of systems thinking. The book also describes how to formally evaluate incident reporting systems to determine whether they meet these requirements.

How to Read This Book

The chapters are designed so that they may be read independently of one another based on the goals of the reader (so there is some repetition of key concepts across chapters). Together, the chapters provide a comprehensive theoretical and practical framework for developing and evaluating an incident reporting system.

The chapters are divided into four main parts:

I. Theoretical framework for designing incident reporting systems
- Chapters 1 and 2 introduce the systems thinking approach, and provide an overview of systems thinking accident causation models and analysis methods. Chapter 1 concludes by highlighting four important principles of the systems thinking approach for the design of incident reporting systems. Chapter 2 presents a case study incident analysis to illustrate the strengths and weaknesses of different systems thinking methods when used as part of an incident reporting system.

- Chapter 3 presents a process model for developing incident reporting systems that is underpinned by the principles of systems thinking, and good practice for injury and incident data collection. The chapter concludes with criteria for designing and evaluating incident reporting systems.

II. Developing and testing incident reporting systems

Chapters 4 to 11 then provide guidance on developing and testing incident reporting systems, using examples from the development of UPLOADS.

- Chapters 4 and 5 describe how to evaluate the state of knowledge on incident causation and data collection in a domain, and identify the priorities of end users prior to designing a new incident reporting system.
- Chapters 6 and 7 describe how to adapt Accimap for use within an incident reporting system, and develop, test, and refine a contributory factor classification scheme.
- Chapters 8, 9, and 10 describe how to develop a prototype incident reporting system, and then test its usability and data quality.
- As an example of the outputs from the previous stages of development, Chapter 11 presents an overview of UPLOADS as implemented in multiple organizations in 2014.

III. Analyzing and using the data from an incident reporting system

Chapters 12 and 13 then provide guidance on analyzing and utilizing the data collected via incident reporting systems underpinned by systems thinking, using examples from the implementation of UPLOADS.

- Chapter 12 describes how to analyze and interpret the data from multiple incident reports.
- Chapter 13 describes how to translate incident data into appropriate, systems thinking–based, incident prevention strategies.

IV. Conclusions and future applications

- The final chapter reflects upon the lessons learned through the development and implementation of UPLOADS, along with some critical future research directions for incident reporting generally.

References

Leveson, N. 2004. A new accident model for engineering safer systems. *Safety Science* 42 (4):237–270. doi: http://dx.doi.org/10.1016/S0925-7535(03)00047-X.

Rasmussen, J. 1997. Risk management in a dynamic society: A modelling problem. *Safety Science* 27 (2/3):183–213. doi: https://doi.org/10.1016/s0925-7535(97)00052-0.

Authors

Natassia Goode is a Senior Research Fellow within the Centre for Human Factors and Sociotechnical Systems at the University of the Sunshine Coast. Dr. Goode currently holds an Advance Queensland Research Fellowship focused on applying systems thinking methods in healthcare. Her PhD and honours research in psychology investigated how people learn about complex systems. Since then, her research has focused on applying systems thinking to optimize the way that organizations manage safety. She has co-authored over 45 peer-reviewed journal articles, and numerous conference contributions and industry reports.

Paul M. Salmon is a Professor in Human Factors and the creator and director of the Centre for Human Factors and Sociotechnical Systems (www.hf-sts.com) at the University of the Sunshine Coast. Professor Salmon currently holds an Australian Research Council Future Fellowship in transport safety. He has a 17-year track record of applied Human Factors research in areas such as road and rail safety, aviation, defence, sport and outdoor recreation, healthcare, workplace safety, land use and urban planning, and cybersecurity. His research has focused on understanding and optimizing human, team, organizational, and system performance through the application of Human Factors theory and methods. He has co-authored 14 books, over 180 peer-reviewed journal articles, and numerous book chapters and conference contributions. He has received many awards and accolades for his research and, in 2016, was awarded the Human Factors and Ergonomics Society Australia's Cumming Memorial Medal for his contribution to Human Factors research and practice.

Michael G. Lenné is an Adjunct Professor (research) at the Monash University Accident Research Centre (MUARC), Victoria, Australia. He obtained a PhD in human factors psychology in 1998 and has since served in a number of research roles in university and government settings. His most recent academic position was Professor in Human Factors at MUARC, where his research program used simulation and instrumented vehicles to study the impacts of vehicle design, technology, and road design on behavior and safety. Professor Lenné's research is widely disseminated, with over 120 journal publications, 5 books, and over 200 reports that provide practical recommendations. He has been the guest editor for three journal special issues since 2014, and served on the editorial boards of the field's leading journals. In 2014, he transitioned into a scientific role with a primary focus on the development of research partnerships that directly support the development of technology to improve transport safety. He maintains academic collaborations through his adjunct role at MUARC, including the flagship Enhanced Crash Investigation Study.

Caroline F. Finch is the Deputy Vice-Chancellor (Research) at the Edith Cowan University in Perth, Australia. Prior to this current role, she held a Robert HT Smith Personal Chair in Sports Safety at Federation University Australia, in Ballarat where she was also supported by a National Health and Medical Research Council Principal Research Fellowship. Since 2010, she has been the Director of the Australian Centre for Research into Injury in Sport and its Prevention (ACRISP), an IOC recognized centre focusing on research in injury prevention and the promotion of health in athletes. Professor Finch is a highly accomplished academic and world-renown researcher. She is the author of over 700 research-related publications, and has earned more than $22 million in research funding over the course of her career. She is known globally for her injury prevention, injury surveillance, and sports medicine research. She has previously been ranked as one of the 10 most highly published injury researchers of all time and is recognized as one of the most influential sports medicine researchers internationally. In 2015, she was awarded the American Public Health Association Distinguished International Career Award from the Injury Control and Emergency Health Services Section. In 2018, Professor Finch was appointed as an Officer of the Order of Australia (AO) for her distinguished service to sports medicine, particularly in the area of injury prevention as an educator, researcher, and author, and to the promotion of improved health in athletes and those who exercise.

Acknowledgements

We acknowledge the many individuals and organizations who have contributed to this program of research over the past 10 years.

It is important to acknowledge that the research was initiated by representatives of the Australian outdoor sector, when they commissioned the Monash University Accident Research Centre to undertake a review of the role of Human Factors in LOA incidents. This initial project led to two successful Australian Research Council Linkage Grants (LP110100037; LP150100148) in partnership with the University of the Sunshine Coast, Federation University Australia, Monash University, the Australian Camps Association, Outdoor Educators' Association of South Australia, Outdoors South Australia, United Church Camping, Outdoors Victoria, Outdoor Council of Australia, Recreation South Australia, Outdoor Recreation Industry Council, Outdoors Western Australia, YMCA Victoria, the Outdoor Education Group, Girl Guides Australia, Wilderness Escape Outdoor Adventures, Venture Corporate Recharge, Queensland Outdoor Recreation Federation, Christian Venues Association, Parks Victoria, Victoria Department of Planning and Community Development, Outdoor Education Australia, and the Department of National Parks, Recreation, Sport and Racing (Queensland).

We thank our project steering committee for their ongoing support, time, and input: Clare Dallat (the Outdoor Education Group), David Strickland (Sport and Recreation Victoria), Brendan Smith (YMCA Victoria), David Petherick and Pete Griffiths (the Australian Camps Association and Outdoor Council of Australia), Andrew Knight and Chuck Berger (Outdoors Victoria), and Andrew Govan (Wilderness Escape Outdoor Adventures and Outdoors South Australia). You have been instrumental in the success and longevity of the project. We especially thank Clare Dallat for the many days she has spent on the road with the research team, co-facilitating industry workshops, and David Strickland for hosting many workshops.

We also acknowledge that this project would not have been possible without the many individuals within the outdoor sector who participated in the research activities, and then implemented the resulting incident reporting system within their organization. This significant, ongoing, contribution of time and effort is greatly appreciated.

We express our appreciation for the contributions of staff and research students at the University of the Sunshine Coast, Federation University Australia, and Monash University who worked on this program of research. We thank Amanda Clacy, Natalie Taylor, Eryn Grant, Erin Stevens, Gemma Read, Michelle van Mulken, Lauren Coventon, Clare Dallat, Tony Carden, Brian Thoroman, Antje Spiertz, Kerri Salmon, Louise Shaw, Nirmala Perera,

Amy Williamson, Eve Mitsopoulos-Rubens, Christina (Missy) Rudin-Brown, Miranda Cornelissen, Margaret Trotter, and Erin Cassell.

Dr. Natassia Goode's contribution was funded through the University of the Sunshine Coast. Professor Paul Salmon's contribution was funded through his ARC Future Fellowship (FT140100681). Professor Caroline Finch was supported by an NHMRC Principal Research Fellowship (ID: 565900).

The studies described in this book were approved by the Monash University or the University of the Sunshine Coast Human Research Ethics Committees.

1

Systems Thinking and Incident Causation

Practitioner Summary

The so-called systems thinking approach is now the dominant approach to understanding and preventing incidents in the safety science literature. As such, there are now various systems thinking–based models of accident causation available, and it is often difficult to determine which is the most appropriate to underpin safety management efforts. This chapter provides an overview of the systems thinking approach, and discusses prominent accident causation models from the contemporary literature. The intention is for the reader to gain an understanding of key systems thinking principles and state-of-the-art accident causation models. In closing, we articulate the key implications of the models for the design of incident reporting systems.

1.1 Introduction

The term 'systems thinking' is used throughout this book to describe a philosophy which can be used to understand and improve performance and safety in complex sociotechnical systems. In human factors and safety science, this philosophy is now widely accepted to be the most appropriate to underpin safety management practices. Accordingly, safety management tools such as accident analysis methods and incident reporting systems should be developed based on systems thinking principles.

One of the key contributions of the systems thinking philosophy is to provide a series of accident causation models that consider incidents as a systems phenomenon. When designing rather than considering incident reporting systems, the importance of using an appropriate accident causation model cannot be understated. The underlying accident causation model determines the type of data collected, the method used to analyze the data, and the recommendations that will be proposed (Underwood & Waterson, 2013).

Notably, our understanding of incident causation has evolved significantly over time, resulting in three distinct types of model: sequential, epidemiological, and systemic (Hollnagel, 2004). Sequential models, such as Heinrich's (1931) domino model, view incidents as a linear sequence of events, with human error and mechanical failures being seen as the primary causes of incidents. Epidemiological models, such as Reason's (1990, 1997) Swiss Cheese model, view incidents causation as similar to the spreading of disease, and emphasize how latent conditions within an organization result in unsafe acts made by operators at the so-called sharp end. Finally, systemic models, such as Rasmussen's (1997) risk management framework, view incidents as the result of multiple decisions and actions across the overall system of work.

Whilst systemic models, or systems thinking–based models as we will call them, are now undoubtedly the most widely accepted, epidemiological models are arguably still dominant in practice (Salmon, Cornelissen, & Trotter, 2012; Underwood & Waterson, 2014). Many researchers have called for better translation of systems thinking–based models in practice. In the case of incident reporting systems, it is our view that systems thinking–based models should provide the underpinning philosophy. The aim of this chapter is therefore to provide an overview of the systems thinking approach, and discuss the most widely applied contemporary models. The intention is for the reader to gain an understanding of key systems thinking principles and state-of-the-art accident causation models. In closing, we articulate the key implications of the models for the design of incident reporting systems.

1.2 Introduction to Systems Thinking

The systems approach to incident causation and analysis is a long and established philosophy that first emerged in the early 1900s (e.g. Heinrich, 1931). It has since evolved through several accident causation models and analysis methods (e.g. Hollnagel, 2012; Leveson, 2004; Perrow, 1984; Rasmussen, 1997; Reason, 1990). Building on key tenets of systems and complexity theory, the philosophy is underpinned by the notion that incidents, and indeed safety, are emergent properties arising from non-linear interactions between components across entire work and social systems (e.g. Leveson, 2004).

There are three key principles of the systems thinking approach that are common across the various accident causation models presented in the literature:

1. *Multiple contributory factors spanning multiple hierarchical system levels.*
   Incidents are created by an interacting web of contributory factors that spans all levels of the work system, from the operational

frontline (worker, equipment, and environment) all the way up to, and including, regulation and government.

2. *Multiple actors and a shared responsibility.* The web of interacting contributory factors is created by the decisions and actions of all actors within the system, including frontline workers, supervisors and managers, chief executives, and government personnel to name only a few. Accordingly, there is a shared responsibility for incident causation and prevention that spans all levels of the work system.

3. *Up and out not down and in.* Incident analysis and prevention efforts should be blame free and take the overall work system as the unit of analysis, rather than the individuals working within it. This involves going 'up and out' rather than 'down and in' during incident analysis. Incident prevention strategies should focus on optimising the interactions between the components in the system, rather than focusing on individual components alone.

The systems thinking approach to safety therefore involves taking the overall system as the unit of analysis. This involves looking beyond the behavior of the individuals involved in incidents and the immediate circumstances of the event. This view also encompasses factors within the broader organizational, social, or political system in which processes or operations take place. Taking a systems thinking perspective, incidents emerge not from the decisions or actions of an individual but from interactions between humans and technology across the wider system. This means that decisions and actions made at government, regulatory, and organizational levels all play a role in incidents. This calls for a more holistic approach to safety management that considers the role of all the actors within the system, as well as the interactions between them.

It is worth noting that the systems thinking approach has only recently emerged as the dominant approach to understanding incident causation. Prior to this, a more deterministic, reductionist approach prevailed (and indeed still does in some areas). This common approach involves going 'down and in' to examine the performance of people and equipment, and their role in incidents. This is reflected in the common fixation on 'human error' and the behavior of frontline operators (e.g. pilots, drivers, control room operators) as the primary cause of incidents. For over four decades, headlines describing 'human error' or behavior as the primary 'root cause' of major incidents have been the norm.

In many organizations, safety management practices are largely driven by this limited perspective. In the case of incident reporting systems, reductionist systems use data fields focused on the behavior of those at the sharp end, not permitting users to report contributory factors related to others across the wider work system. Consequently, the resulting prevention strategies tend to focus predominantly on frontline operators and aim to improve their behavior through education, training, enforcement, and the prohibition of undesirable behaviors. Whilst in some cases parts of safety-critical systems may be improved (e.g. through training programs), little consideration is given to how these parts interact with one another, or how the system functions as a whole.

1.3 Systems Thinking Applied

To demonstrate the differences between the systems thinking approach, and the reductionist human error approach, it is worth discussing both perspectives when applied to a recent high-profile incident. For this purpose, we use the tragic Air France crash of 2009, in which an Airbus A330 stalled and crashed into the Atlantic Ocean, killing all 228 people on board.

The incident occurred on the 31st May 2009 during a scheduled passenger flight from Rio de Janeiro, Brazil, to Paris, France. Just over three and a half hours after departure, the captain left the flight deck to take a rest break. Following this, the pilot flying (PF) noted that the plane had entered the Intertropical Convergence Zone (TCZ), which is an area close to the equator that experiences severe weather conditions. Shortly afterwards, the PF called through to a flight attendant to warn of impending turbulence and the need to take care.

Upon entry into the TCZ, the aircraft's pitot tubes (the devices which measure airspeed) froze due to the low air temperature. As a result, the data being sent to the cockpit was incorrect, which in turn caused the autopilot to disconnect (with an accompanying cockpit alert sounding to notify the pilots). Shortly after, the PF remarked, 'I have the controls' and was acknowledged by the pilot not flying (PNF). The PF then put the aeroplane into a steep climb by pulling back on his sidestick. The plane gained altitude rapidly but lost speed quickly, triggering a stall warning that subsequently sounded 75 times. The PF continued to apply nose up inputs with the PNF apparently unaware of this. The aeroplane went into a stall and began to lose altitude. Unable to diagnose the situation, the PNF called the captain back into the cockpit. The PF announced, 'I don't have control of the plane', following which the PNF took control of the aeroplane. Six seconds later, the captain returned to the cockpit and joined in the effort to diagnose the situation. Both the PF and PNF informed the captain that they had lost control of the aircraft and did not understand why. The PF told the captain that he had 'had the stick back the whole time', at which point the PNF took control of the plane and applied nose down inputs to prevent the stall and gain speed. This action was taken too late, and the plane crashed into the ocean, killing all onboard.

The tragic Air France incident illustrates the juxtaposition between the reductionist human error approach and the systems thinking approach to incident analysis and prevention. A reductionist approach would go 'down and in' and focus primarily on the aircrew's role in the incident, seeking to identify errors or mistakes that led to the stall and crash. For example, the Bureau d'Enquetes et d'Analyses's (BEA) investigation report describes the aircrew's inability to understand and respond to the situation that arose following the freezing of the pitot tubes, whereby the autopilot disconnected them, and the plane required manual control. Following some confusion, the pilots were seemingly unaware that the plane had entered a stall and was descending rapidly towards the ocean. Specifically, the BEA reported that the crew failed to make a link between the loss of indicated airspeeds and the appropriate procedure, made a late identification of the deviation from flight path, and failed to identify the approach to stall and the subsequent stall situation itself (Bureau d'Enquêtes et d'Analyses, 2012). A reductionist approach would stop here, with the inevitable conclusion that the primary causes of the incident were the loss of situation awareness by the aircrew and the subsequent erroneous control inputs made by the PF.

The systems thinking approach, however, would continue to go 'up and out' and look at the factors across the aviation system that influenced the aircrew's behavior. This involves looking at interactions between the components within the system; a key aspect being the interactions between the human and non-human agents. For example, the incident was initiated through an interaction between two non-human agents, the pitot tubes and the autopilot system, resulting in the autopilot disconnection. A systems thinking–based investigation would also look beyond the immediate environment (i.e. cockpit), and the events immediately prior to the incident. This would include examining the airspace, the relevant air traffic controllers, design standards, maintenance, regulation and the aircrew's experience and training. The conclusions from this approach reveals an interacting web of contributory factors related to the pilot, the aircrew, the captain, the cockpit, the aircraft, the flight environment, air traffic controllers, the airline company's policy, procedures and training, aircraft designers, regulatory bodies, government, and so on. Indeed, the Bureau d'Enquetes et d'Analyses's (BEA) investigation found issues with the training received by the pilots, the design of the cockpit, the air traffic control system, and the aircrew's planning and preparation for crossing the intertropical convergence zone.

The ensuing response in both cases would be very different. In the former scenario, the pilot is blamed and, if alive, would likely suffer reprisals. Subsequent prevention strategies would likely focus on improvements to pilot training designed to ensure that the same errors would not be made on future flights. From the systems thinking approach, the investigation body would seek to drive fundamental and broader changes to the aviation system to ensure that similar events could not occur again.

1.4 Systems Thinking Models

As mentioned above, various systems thinking–based models of accident causation are presented in the literature (e.g. Dekker, 2011; Hollnagel, 2012; Leveson, 2004; Perrow, 1984; Rasmussen, 1997). For the purposes of this introductory chapter, we describe the three that are arguably the most closely aligned with systems theory: Rasmussen's risk management framework (RMF), Leveson's Systems Theoretic Accident Model and Processes (STAMP; Leveson, 2004), and Dekker's Drift Into Failure (DIF) model (Dekker, 2011).

1.4.1 Rasmussen's Risk Management Framework

Rasmussen's risk management framework (Rasmussen, 1997; see Figure 1.1) is arguably the most popular contemporary systems thinking accident causation model (Salmon et al., 2017; Waterson, Jenkins, Salmon, & Underwood, 2017). Indeed, to date, the model has been used to support analyses of incidents and safety issues in a wide range of domains, including outdoor recreation (Salmon et al., 2017), freight handling (Goode, Salmon, Lenné, & Hillard, 2014), maritime (Kee, Jun, Waterson, & Haslam, 2017), rail (Salmon, Read, Stanton, & Lenné, 2013), emergency response (Jenkins, Salmon, Stanton, & Walker, 2010), road safety (Newnam & Goode, 2015), *E. coli* outbreaks (Nayak & Waterson, 2016), healthcare (Waterson, 2009), and process control (Fabiano, Vianello, Reverberi, Lunghi, & Maschio, 2017).

The framework is underpinned by the following ideas:

- Systems are comprised of various hierarchical levels (e.g. government, regulators, company, company management, staff, and work), each of which contain actors (individuals, organizations, or technologies) who share responsibility for production and safety;
- Decisions and actions occurring at all levels of the system interact to shape system performance, meaning both safety and incidents are influenced by the decisions of all actors, not just frontline workers; and
- Incidents are caused by multiple, interacting factors, not just one bad decision or single action.

A key implication is that it is not possible to truly understand safety and performance by decomposing the system and examining its components alone; rather, it is the interactions between the components that are important. Further, the more components and interactions studied (i.e. the higher

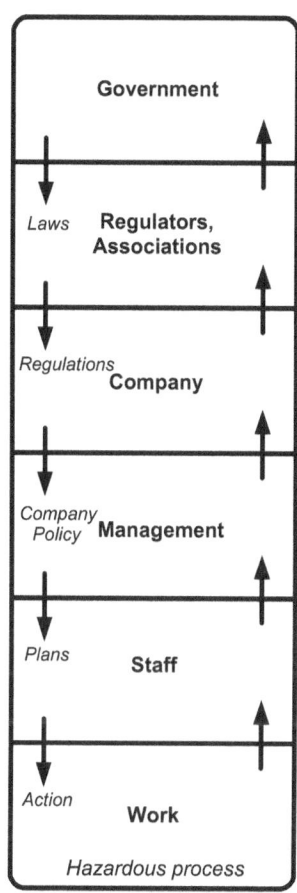

FIGURE 1.1
Rasmussen's risk management framework. (From Rasmussen, J. 1997. Risk management in a dynamic society: A modelling problem. *Safety Science, 27*(2–3), 183–213. doi:10.1016/S0925-7535(97)00052-0. With permission.)

up into the system analyses go), the closer one can get to understanding system performance and the factors influencing it.

A notable feature of Rasmussen's framework (and other systems thinking models) is the notion that the behaviors underpinning incidents do not necessarily have to be errors, failures, or violations (e.g. Dekker, 2011; Leveson, 2004; Rasmussen, 1997). As Dekker (2011) points out, systems thinking is about how incidents can happen when no parts are broken. Normal performance plays a role too (Perrow, 1984). Indeed, it is argued that incidents arise from the very same behaviors and processes that create safety. These normal behaviors include workarounds, improvisations, and adaptations (Dekker, 2011), but may also just be normal work behaviors routinely undertaken to get the job done, or the design of management systems (e.g. rostering, work scheduling). It is only with hindsight that these normal behaviors and conditions of work are treated as failures during incident investigations.

Actors at each level of Rasmussen's framework are involved in safety and performance management. For systems to function efficiently and safely, decisions made at higher governmental, regulatory, and managerial levels of the system should propagate down, and be reflected in the decisions and actions occurring at the lower levels. Conversely, information at the lower levels regarding the system status needs to transfer up the hierarchy to inform the decisions and actions occurring at the higher levels (Cassano-Piche, Vicente, & Jamieson, 2009). Without this so called 'vertical integration', systems lose control of the hazardous processes that they are designed to control (Cassano-Piche et al., 2009).

Rasmussen also outlined the concept of migration, describing how organizations drift toward, and away from, safety boundaries due to various constraints, including financial, production, and performance pressures (see Figure 1.2). According to Rasmussen, there is a boundary of economic failure: these are the financial constraints on a system that influence behavior towards greater cost efficiencies. There is also a boundary of unacceptable workload: these are the pressures experienced by people and

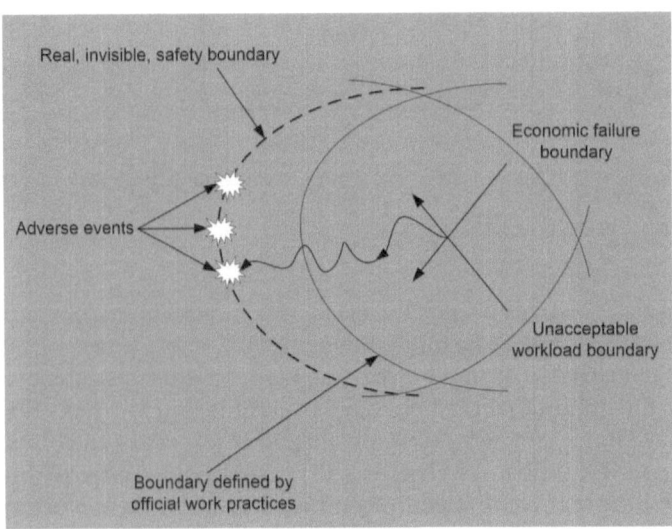

FIGURE 1.2
Rasmussen's dynamic safety space. (From Rasmussen, J. 1997. Risk management in a dynamic society: A modelling problem. *Safety Science*, 27(2–3), 183–213. doi:10.1016/S0925-7535(97)00052-0. With permission.)

equipment in the system as they try to meet economic and financial objectives. The boundary of economic failure creates a pressure towards greater efficiency, which works in opposition to excessive workload. As systems involve human as well as technical elements, and because humans are able to adapt situations to suit their own needs and preferences, these pressures inevitably introduce variations in behavior that are not explicitly designed, and can lead to increasingly emergent system behaviors, both good and bad (Clegg, 2000; Qureshi, 2007). Over time, this adaptive behavior can cause a system to cross safety boundaries, and incidents to happen (Qureshi, 2007; Rasmussen, 1997). The key, then, is to detect in advance (a) where those boundaries are; and (b) where the system is travelling in relation to them.

To summarize, Rasmussen's framework makes a series of assertions that provide a framework for understanding incident causation in many domains. These so-called tenets are summarized as follows:

1. Incidents are emergent properties impacted by the decisions and actions of all actors, not just frontline operators alone;
2. Threats to safety are caused by multiple contributing factors, not just a single poor decision or action;
3. Threats to safety can result from a lack of vertical integration across levels of the work system, not just from deficiencies at one level alone;
4. Lack of vertical integration is caused, in part, by a lack of feedback across levels of the work system; and
5. System behaviors are not static – they migrate over time and under the influence of various pressures, such as financial and psychological pressures;
6. Migration occurs at multiple levels of the system;
7. Migration of practices cause system defences to degrade and erode gradually over time, not all at once. Incidents are caused by a combination of this migration and a triggering event(s).

1.4.2 Systems Theoretic Accident Model and Processes (STAMP)

Inspired by Rasmussen's risk management framework, Leveson introduced the STAMP model and associated accident analysis method in 2004. Underpinned by systems and control theory, STAMP views safety and incidents primarily as control problems. Specifically, Leveson takes the view that incidents occur when safety controls are either inadequate, not enforced, or not in place.

A broad view of controls is adopted by the model. Leveson (2004) describes how behavior is controlled not only by engineered systems and direct intervention, but also by policies, procedures, shared values, and other aspects of the organizational and social culture (Leveson, 2011). In relation to the Air France crash described earlier, for example, STAMP would suggest that the crash occurred because the system of controls designed to prevent a stall mid-flight failed. These controls would include pilot training, cockpit systems such as stall warnings and displays, the autopilot system, flight laws, procedures, threat and error management, air traffic control, and regulatory frameworks to name only a few.

Leveson (2004) outlines a general control structure to demonstrate how work systems manage safety (see Figure 1.3). The left-hand side of Figure 1.3 shows a generic control structure for system development, whereas the right-hand side shows a generic control structure for system operation. The STAMP control structure model views systems as comprising many inter-related components (e.g. actors, organizations, equipment, procedures, rules and regulations) that are kept safe through control and feedback loops (Leveson, 2004). Accordingly, control structure models incorporate a series of hierarchical system levels and describe the actors and organizations that reside at each level. Control and feedback loops are included to show what control mechanisms are enacted down the hierarchy, and what information about the status of the system is sent back up the hierarchy. The arrows flowing down the hierarchy represent control relationships (or reference channels), and the arrows flowing up the hierarchy represent feedback loops (or measuring channels). In relation to the Air France example described earlier, the airline company would enact the control of 'pilot training' on their pilots. In turn, data on pilots' performance in the training programs, and during actual flights, would represent feedback mechanisms to the training and safety managers within the airline. Similar to Rasmussen's concept of vertical integration, this feedback would allow the airline to update its training programs to deal with new and emergent issues on the flight deck.

1.4.3 Drift into Failure (DIF) Model

Building on systems models such as Rasmussen's and Leveson's, as well as key concepts from complexity science, Dekker's (2011) DIF model describes how complex system performance gradually shifts, often unchecked and unrecognized, towards incidents. According to the DIF model, incidents result from the non-linear interactions between what often seem locally to be very normal behaviors. Dekker (2011) argues, for example, that the seeds for failure can be found in 'normal, day-to-day processes' (pg. 99), and are often driven by goal conflicts and production pressures. Multiple decisions and actions, occurring over time, in different contexts, under different constraints, and with only limited knowledge of their effects, gradually lead a system to adapt in unforeseen ways. Ultimately, these adaptations can produce incidents.

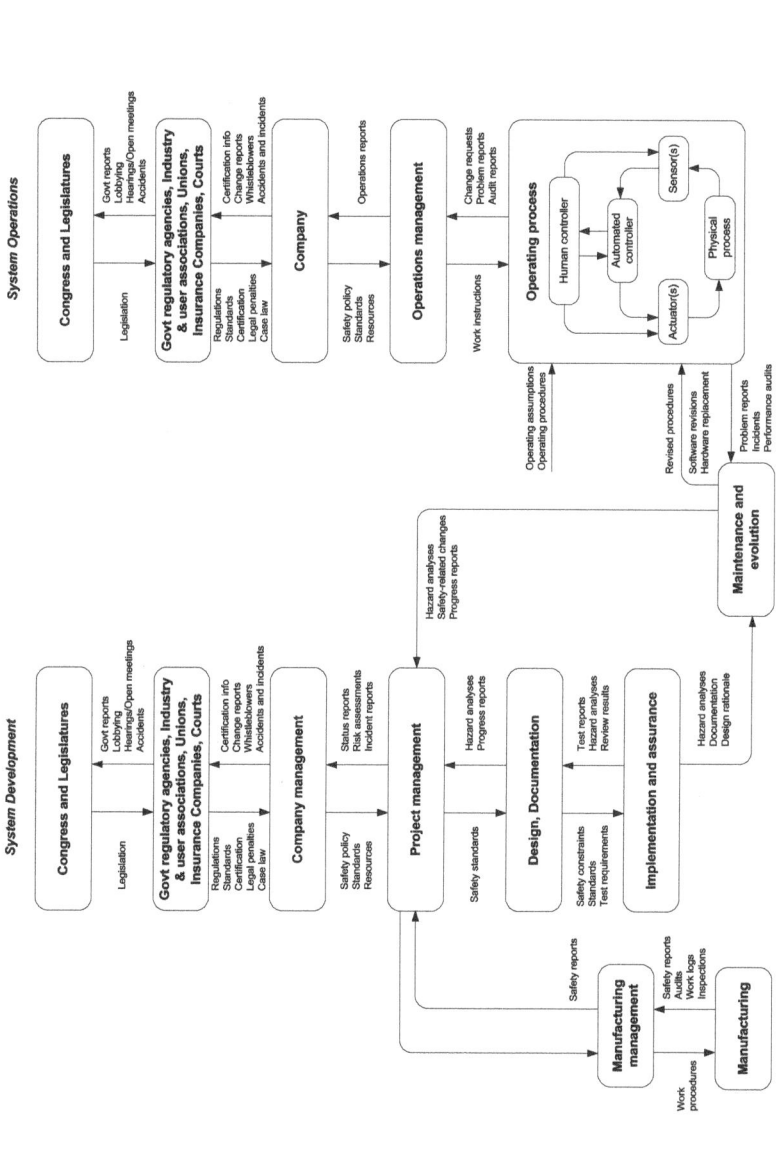

FIGURE 1.3

STAMP generic control structure model. (From Leveson, N. G. 2004. A new accident model for engineering safer systems. *Safety Science, 42*(4), 237–270. doi:10.1016/S0925-7535(03)00047-X. With permission.)

The DIF model outlines five key characteristics of drift: scarcity and competition, decrementalism, sensitive dependence on initial conditions, unruly technologies, and the contribution of protective structures. Scarcity and competition reflects the limited availability of resources, and the competition between organizations present in most complex systems. As a result, multiple trade-offs are made in order to remove, or balance, resource limitations and production pressures; often this leads to a steady adaptation of processes and technologies that lean toward unsafe practices (Dekker, 2011).

Decrementalism refers to the gradual, step-by-step slipping of safe practice to unsafe practice. Here, continuous minute adaptations to practice driven by different system components operating under various constraints, eventually lead to large failures (Dekker, 2011). Each small adaptation is accepted, as it is only a minor departure from the previously accepted norm, and subsequent successful performance is taken as an indicator that the adaptation does not impact safety (Dekker, 2011). In reality, each small adaptation is another step away from safe practices towards unsafe ones.

Sensitive dependence on initial conditions refers to the notion that even miniscule changes in initial conditions can lead to dramatic changes in system behavior (Hilborn, 2004). This essentially means that small decisions or actions made some time ago, during the design of a work system, can interact with other components, to create catastrophic events down the track.

Unruly technology reflects the lack of control that system designers and managers have over the technologies that they introduce. Despite the best efforts of designers, certification, regulators, and so on, new technologies often behave in unexpected ways when introduced into complex systems (Dekker, 2011). This is due to unforeseen interactions with other components, variance in how the technologies are used, and unforeseen contexts in which the technologies are used.

Finally, contribution of protective structures refers to the many disconnected structures that are designed to ensure that systems operate safely. These include regulatory arrangements, safety committees and teams, and quality review and certification boards to name only a few. According to Dekker (2011), in addition to failing to intervene when they should, protective structures can actively contribute to drift through poor knowledge, lack of access and information, conflicting goals, and decisions that make only local sense. Control measures brought in to solve one problem can introduce new problems, pushing the system towards an incident.

1.4.4 A Note on Other Accident Causation Models

In this chapter, we have purposely focused on three state-of-the-art systems thinking models of accident causation. It is worth noting that other models exist, including Reason's Swiss Cheese model (Reason, 1990, 1997), Hollnagel's Functional Resonance Analysis Method (FRAM; Hollnagel, 2012), and Perrow's Normal Accident Theory (NAT; Perrow, 1984). Of these, Reason's Swiss Cheese model is the most popular, and indeed is arguably still the most commonly applied model in practice. Its popularity is such that it has driven the development of various accident analysis methods, for example the aviation accident analysis method, Human Factors Analysis and Classification System (HFACS) (Wiegmann & Shappell, 2003). The Swiss Cheese model has also been commonly applied as an accident analysis framework itself (e.g. Lawton & Ward, 2005).

Reason's Swiss Cheese model describes the interaction between system wide 'latent conditions' (e.g. poor designs and inadequate equipment, inadequate supervision, manufacturing defects, maintenance failures, inadequate training, poor procedures) and unsafe acts made by human operators. The model describes how these latent conditions reside across all levels of the organizational system (e.g. higher supervisory and managerial levels). According to the model, each organizational level has defences, such as protective equipment, rules and regulations, training, checklists and engineered safety features, which are designed to prevent the occurrence of incidents. Weaknesses in these defences, created by latent conditions and unsafe acts, create 'windows of opportunity' for incident trajectories to breach the defences and cause an incident.

There is no doubt that Reason's Swiss Cheese model has shaped the face of incident analysis and prevention research and practice for well over two decades. However, it is the authors' opinion that it is not suitable to provide the underpinning framework for an incident reporting system. Whilst the model has made a significant contribution to the development of safety management systems, it is now outdated and does not fully align with key systems thinking principles. In its original form it is focused on contributory factors within a particular organization and does not consider wider systemic influences, such as government regulation. In addition, although it considers how latent conditions might create unsafe acts, it does not explicitly consider the interactions between them.

1.5 Implications for Incident and Reporting Systems

The systems thinking–based models of accident causation described above have several important implications for the design of incident reporting systems:

1. *'Up and out' rather than 'down and in'.* To be effective, incident reporting systems need to collect data on contributory factors from across the overall work system, in addition to factors relating to the immediate context of the event (e.g. frontline workers, environment, and

equipment). Even when it is clear that individual behavior, such as pilot error, driver distraction, or impairment, was involved in an incident, there are still many other interacting factors that explain why the system allowed that behavior to occur. In short, incident reporting systems should encourage users to 'look up and out' and describe the network of contributory factors involved, rather than 'down and in' on the behavior of frontline operators.

2. *It's all about interactions and relationships.* Incident reporting systems should encourage users to provide data regarding the interactions, or relationships, between contributory factors both within, and across, system levels. Alternatively, the data reported should support identification of these interactions and relationships. A key principle of all systems thinking accident causation models is that incidents are emergent properties arising from interactions between system components. Incident reporting systems must be capable of capturing these interactions and resulting emergent behaviors.

3. *Avoid focusing only on failure.* Incident reporting systems should be blame free and should not focus exclusively on failures (e.g. errors, violations, lapses) or 'gaps' in defences. Systems thinking models are underpinned by the idea that incidents can happen when no parts are broken. Incident reporting systems should therefore enable users to describe how the normal conditions of work contributed to incidents.

4. *A systems lens.* Finally, there is a need to provide practitioners with an appropriate methodological framework with which to report, analyze and represent incidents from a systems perspective. As incident reporters are often involved in the event, it is unlikely that they will automatically consider incident causation from a systems thinking perspective. Appropriate training and prompts need to be provided for incident reporters to understand, and adopt, this perspective. As a corollary, the systems approach also needs to be firmly embedded in the analysis and interpretation of the incident data that is collected.

References

Bureau d'Enquêtes et d'Analyses. (2012). *Final report on the accident on 1st June 2009 to the Airbus A330-203 registered F-GZCP operated by Air France flight AF 447 Rio de Janeiro–Paris*. Retrieved from http://www.bea.aero/docspa/2009/f-cp090601 .en/pdf/f-cp090601.en.pdf, accessed 18th June 2015.

Cassano-Piche, A. L., Vicente, K. J., & Jamieson, G. A. (2009). A test of Rasmussen's risk management framework in the food safety domain: BSE in the UK. *Theoretical Issues in Ergonomics Science, 10*(4), 283–304. doi:10.1080/14639220802059232

Clegg, C. W. (2000). Sociotechnical principles for system design. *Applied Ergonomics, 31*(5), 463–477. doi:10.1016/S0003-6870(00)00009-0

Dekker, S. (2011). *Drift into failure: From hunting broken components to understanding complex systems.* Aldershot, UK: Ashgate.

Fabiano, B., Vianello, C., Reverberi, A. P., Lunghi, E., & Maschio, G. (2017). A perspective on Seveso accident based on cause-consequences analysis by three different methods. *Journal of Loss Prevention in the Process Industries, 49,* 18–35. doi:10.1016/j.jlp.2017.01.021

Goode, N., Salmon, P. M., Lenné, M. G., & Hillard, P. (2014). Systems thinking applied to safety during manual handling tasks in the transport and storage industry. *Accident Analysis & Prevention, 68,* 181–191. doi:10.1016/j.aap.2013.09.025

Heinrich, H. W. (1931). *Industrial accident prevention: A scientific approach.* New York, McGraw-Hill.

Hilborn, R. C. (2004). Sea gulls, butterflies, and grasshoppers: A brief history of the butterfly effect in nonlinear dynamics. *American Journal of Physics, 72*(4), 425–427. doi:10.1119/1.1636492

Hollnagel, E. (2004). *Barriers and accident prevention.* Aldershot, UK: Ashgate.

Hollnagel, E. (2012). *FRAM, the functional resonance analysis method: Modelling complex socio-technical systems.* Aldershot, UK: Ashgate.

Jenkins, D. P., Salmon, P. M., Stanton, N. A., & Walker, G. H. (2010). A systemic approach to accident analysis: A case study of the Stockwell shooting. *Ergonomics, 53*(1), 1–17. doi:10.1080/00140130903311625

Kee, D., Jun, G. T., Waterson, P., & Haslam, R. (2017). A systemic analysis of South Korea Sewol ferry accident – Striking a balance between learning and accountability. *Applied Ergonomics, 59,* 504–516. doi:10.1016/j.apergo.2016.07.014

Lawton, R., & Ward, N. J. (2005). A systems analysis of the Ladbroke Grove rail crash. *Accident Analysis & Prevention, 37*(2), 235–244. doi:10.1016/j.aap.2004.08.001

Leveson, N. G. (2004). A new accident model for engineering safer systems. *Safety Science, 42*(4), 237–270. doi:10.1016/S0925-7535(03)00047-X

Leveson, N. G. (2011). Applying systems thinking to analyze and learn from events. *Safety Science, 49*(1), 55–64. doi:10.1016/j.ssci.2009.12.021

Nayak, R., & Waterson, P. (2016). 'When Food Kills': A socio-technical systems analysis of the UK Pennington 1996 and 2005 *E. coli* O157 Outbreak reports. *Safety Science, 86,* 36–47. doi:10.1016/j.ssci.2016.02.007

Newnam, S., & Goode, N. (2015). Do not blame the driver: A systems analysis of the causes of road freight crashes. *Accident Analysis & Prevention, 76,* 141–151. doi:10.1016/j.aap.2015.01.016

Perrow, C. (1984). *Normal accidents: Living with high risk technologies.* New York: Basic Books.

Qureshi, Z. H. (2007). *A review of accident modelling approaches for complex socio-technical systems.* Paper presented at the Conferences in Research and Practice in Information Technology Series.

Rasmussen, J. (1997). Risk management in a dynamic society: A modelling problem. *Safety Science, 27*(2–3), 183–213. doi:10.1016/S0925-7535(97)00052-0

Reason, J. (1990). *Human error:* Cambridge, UK: Cambridge University Press.

Reason, J. (1997). *Managing the risks of organizational accidents.* Aldershot, UK: Ashgate.

Salmon, P. M., Cornelissen, M., & Trotter, M. J. (2012). Systems-based accident analysis methods: A comparison of Accimap, HFACS, and STAMP. *Safety Science, 50*(4), 1158–1170. doi:10.1016/j.ssci.2011.11.009

Salmon, P. M., Goode, N., Taylor, N., Lenné, M. G., Dallat, C. E., & Finch, C. F. (2017). Rasmussen's legacy in the great outdoors: A new incident reporting and learning system for led outdoor activities. *Applied Ergonomics, 59*, 637–648. doi:10.1016/j .apergo.2015.07.017

Salmon, P. M., Read, G., Stanton, N. A., & Lenné, M. G. (2013). The crash at Kerang: Investigating systemic and psychological factors leading to unintentional non-compliance at rail level crossings. *Accident Analysis & Prevention, 50*, 1278–1288. doi:10.1016/j.aap.2012.09.029

Underwood, P., & Waterson, P. (2013). *Accident analysis models and methods: Guidance for safety professionals.* UK: Loughborough University.

Underwood, P., & Waterson, P. (2014). Systems thinking, the Swiss Cheese model and accident analysis: A comparative systemic analysis of the Grayrigg train derailment using the ATSB, AcciMap and STAMP models. *Accident Analysis & Prevention, 68*, 75–94. doi:10.1016/j.aap.2013.07.027

Waterson, P. (2009). A systems ergonomics analysis of the Maidstone and Tunbridge Wells infection outbreaks. *Ergonomics, 52*(10), 1196–1205. doi:10.1080 /00140130903045629

Waterson, P., Jenkins, D. P., Salmon, P. M., & Underwood, P. (2017). 'Remixing Rasmussen': The evolution of Accimaps within systemic accident analysis. *Applied Ergonomics, 59*, 483–503. doi:10.1016/j.apergo.2016.09.004

Wiegmann, D. A., & Shappell, S. A. (2003). *A human error approach to aviation accident analysis: The human factors analysis and classification system.* Aldershot, UK: Ashgate.

2

Systems Thinking and Incident Analysis

Practitioner Summary

Methods underpinned by systems thinking are now the dominant approach to incident analysis. In particular, three distinct methods are currently popular in both research and practice: Accimap (Rasmussen, 1997), Causal Analysis using System Theory (CAST; Leveson, 2004), and the Human Factors Analysis and Classification System (HFACS; Wiegmann & Shappell, 2003). All three could potentially form part of an organization's incident reporting system. This chapter provides an overview of each method, and examines their main strengths and weaknesses in light of the incident reporting system requirements identified in Chapter 1. The outputs derived from each method are demonstrated using a case study analysis of the Mangatepopo Gorge walking tragedy. Based on considerations of comprehensiveness, ease of use, flexibility, and utility of outputs, it is argued that Accimap, with a classification scheme of contributory factors, is the most suitable approach for use as part of an incident reporting system.

2.1 Introduction to Incident Analysis

Incident analysis involves using a structured methodology to describe and understand the contributory factors involved in incidents. The methodology that analysts use to describe and analyze incidents is critical, as it determines the types of contributory factors that will be identified, as well as the recommendations that will be proposed to improve safety. If a structured approach to analyzing incidents is not employed, then the analysis will be driven by people's assumptions, and biases, about the causes of the incident. This typically leads people towards finding something, or someone, to blame. Without appropriate methods, our understanding of incidents and our ability to prevent them is limited.

Numerous methods have been developed to analyse incidents (Salmon et al., 2011). In recent times, methods underpinned by systems thinking have become dominant (Salmon et al., 2011). In particular, methods such as Accimap (Rasmussen, 1997; Svedung & Rasmussen, 2002), the Human Factors Analysis and Classification System (HFACS; Wiegmann & Shappell, 2003), and Causal Analysis using System Theory (CAST; Leveson, 2004) are being increasingly applied in both research and practice. Each provides its own distinct approach to incident analysis, and likewise the outputs produced are markedly different. Accimap is a generic approach that is used to identify and link contributory factors across the six system levels described by Rasmussen's risk management framework. The CAST method uses a control structure to identify where controls and feedback mechanisms failed. Finally, HFACS uses a scheme to classify errors, violations and performance shaping factors across the organizational levels specified by Reason's Swiss Cheese model. Although all three are popular, there are significant differences in terms of theoretical underpinning, the methodological approach adopted, the outputs produced, and subsequently the recommendations that might be identified to improve safety.

This chapter provides an overview of each method and compares and contrasts them, via a case study, to identify their main strengths and weaknesses when used as part of an organization's incident reporting system.

2.2 Accimap

Accimap provides a framework, as shown in Figure 2.1, with which to describe incidents through the lens of Rasmussen's risk management framework (Rasmussen, 1997; see Chapter 1).

Specifically, Accimap is used to describe incidents in terms of the contributory factors involved and the relationships between them. It does this by decomposing systems into six hierarchical levels, across which analysts place the contributory factors that enabled the event in question to occur (although the method is flexible in that the number of levels and their labels can be adapted to reflect the structure of the system under analysis). Interactions between these factors are subsequently mapped onto the diagram to show the relationships within and across the six levels shown in Figure 2.1. A notable feature of Accimap is that it does not provide analysts with a scheme or taxonomy for identifying or classifying contributing factor. Rather, analysts have the freedom to incorporate any factor deemed to have played a role in the incident in question.

The method was developed on the basis that task analytic methods are inadequate for analyzing incidents as they are not capable of modelling the

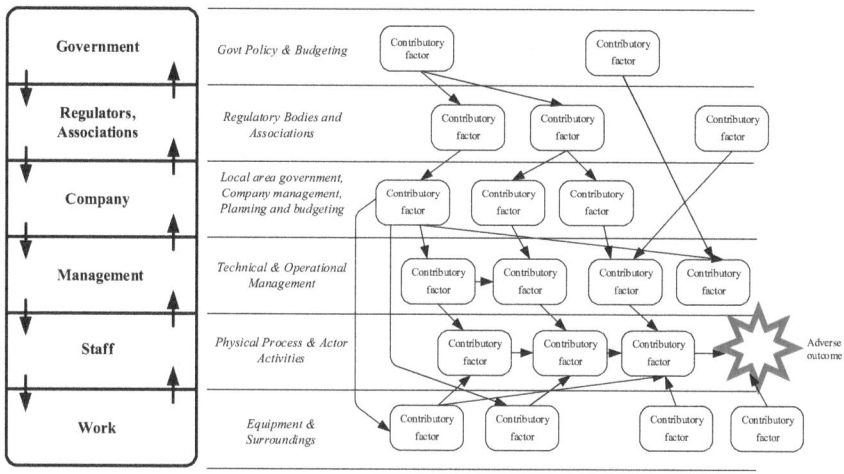

FIGURE 2.1
Rasmussen's risk management framework and Accimap.

systemic network of factors involved. As it is based on Rasmussen's risk management framework, Accimap is also underpinned by the idea that behavior, safety, and incidents are emergent properties of complex socio-technical systems. This means that they are created by the interactions between people, activities, and artefacts across the system, and the decisions and actions of all actors in the system – politicians, chief executives, managers, safety officers, and work planners – not just by frontline workers alone (Cassano-Piche, Vicente, & Jamieson, 2009). To ensure analysts focus on the decisions and actors of all actors in the system, Accimap typically uses the following levels:

- Government policy and budgeting;
- Regulatory bodies and associations;
- Local area government, Company management, Planning and budgeting;
- Technical and operational management;
- Physical processes and actor activities; and
- Equipment and surroundings.

Contributory factors at each of the levels are identified and linked between and across levels based on cause-effect relations (see Figure 2.1). It is important to note that Accimap considers contributory factors across the entire work system (up to and including government), and that it describes the interactions and relationships between contributory factors. This allows the causal flow of events across entire work systems to be identified, which enables identification of the planning, management, regulatory and governing bodies that may have made a contribution (Svedung & Rasmussen, 2002). This is particularly powerful for the development of strategies to improve system performance and safety.

2.3 CAST

Although also based on Rasmussen's risk management framework, the CAST method focusses explicitly on control and feedback mechanisms, and takes the view that incidents emerge when disturbances, failures, or dysfunctional interactions between components are not handled by control mechanisms (Leveson, 2004). As discussed in Chapter 1, the underpinning accident causation model, STAMP, views safety as an issue of control that is managed through a control structure that has the goal of enforcing constraints on actors across the system. Leveson (2004) takes a broad view of controls, suggesting that they include not only managerial, organizational, physical, operational, and manufacturing-based controls, but also individual interests, shared values, organizational contexts, and social controls.

Applying CAST to an incident involves: developing a control structure for the system in which the incident occurred; classifying the control and feedback failures involved in the incident; and examining the contextual factors and process models of relevant actors within the system to identify whether inaccurate or incongruent process models contributed to the incident.

A generic control structure, including both system design (development) and system operations, is presented in Figure 2.2. The generic structure is first modified to reflect the system in which the incident occurred. The model incorporates a series of hierarchical system levels, similar to Accimap. Developing a control structure for the analysis of a particular incident involves: (a) adapting the levels of the control structure to describe the specific system in question; and (b) identifying the actors and organizations that reside at each level. Control and feedback loops are included to show what control mechanisms are enacted down the hierarchy, and what system status information is sent back up the hierarchy.

Within Figure 2.2, downward pointing arrows and the associated text represent the control mechanisms imposed by actors and/or organizations on actors and/or organizations at the level below. For example, in a typical work system, supervisors and managers at the operations management level impose controls on workers at the operating process level via standard operating procedures and working rules. The arrows pointing upwards represent feedback mechanisms whereby actors and organizations provide

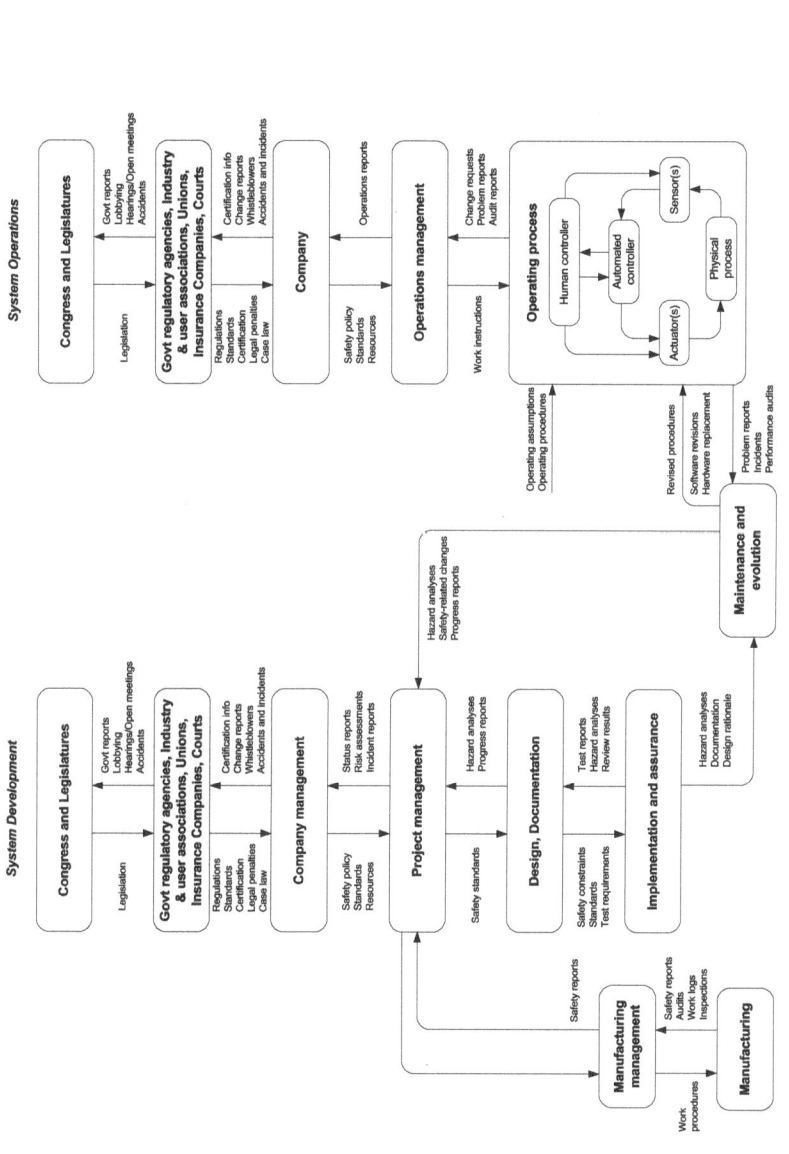

FIGURE 2.2

Generic control structures for system design and system operations. (From Leveson, N. G. 2004. A new accident model for engineering safer systems. *Safety Science, 42*(4), 237–270. doi:10.1016/S0925-7535(03)00047-X. With permission.)

Inadequate enforcement of constraints
- Unidentified hazards - Inappropriate, ineffective, or missing control actions - Failure of control process to enforce constraints - Inconsistent/incomplete/incorrect process models - Inadequate coordination btw controllers and decision makers
Inadequate execution of control actions
- Communications failures - Inadequate actuator operations - Time lag
Feedback failures
- Not provided through system design - Communications failures - Time lag - Inadequate sensor operation

FIGURE 2.3
STAMP control and feedback failure taxonomy. (From Leveson, N. G. 2004. A new accident model for engineering safer systems. *Safety Science*, 42(4), 237–270. doi:10.1016/S0925-7535(03)00047-X. With permission.)

information regarding the status of the system to the levels above. For example, incident reports are a feedback mechanism from the operating process level to the operations management level, which in turn supports management decision making around safety management. Together, these control and feedback mechanisms are used to achieve vertical integration (described in Chapter 1).

A taxonomy of control and feedback failures is then used to support the identification of control and feedback failures that contributed to the incident. The taxonomy of control and feedback failures is presented in Figure 2.3.

2.4 HFACS

HFACS (Wiegmann & Shappell, 2003) was originally developed for use in the analysis of civil and military aviation incidents. The impetus came from the absence of a classification scheme of latent failures and unsafe acts to support the use of Reason's Swiss Cheese model during incident analysis. Accordingly, Wiegmann and Shappell (2003) set out to develop detailed classifications for each level of the model to support its application to aviation incidents. HFACS was subsequently developed based on Reason's model and an analysis of aviation incident reports (Wiegmann & Shappell, 2003). The method itself provides a scheme for classifying errors, violations, and performance shaping factors at each of the following four levels:

- Unsafe acts;
- Preconditions for unsafe acts;
- Unsafe supervision; and
- Organizational influences.

The structure of the HFACS method is presented in Figure 2.4, which shows the higher level categories mapped onto Reason's Swiss Cheese model. Working backward from the immediate unsafe acts, analysts use the scheme to classify the errors and associated contributory factors involved in an incident.

As an example of the level of detail within the scheme, the unsafe acts categories are presented in Table 2.1. The complete scheme can be found in Wiegmann and Shappell (2003).

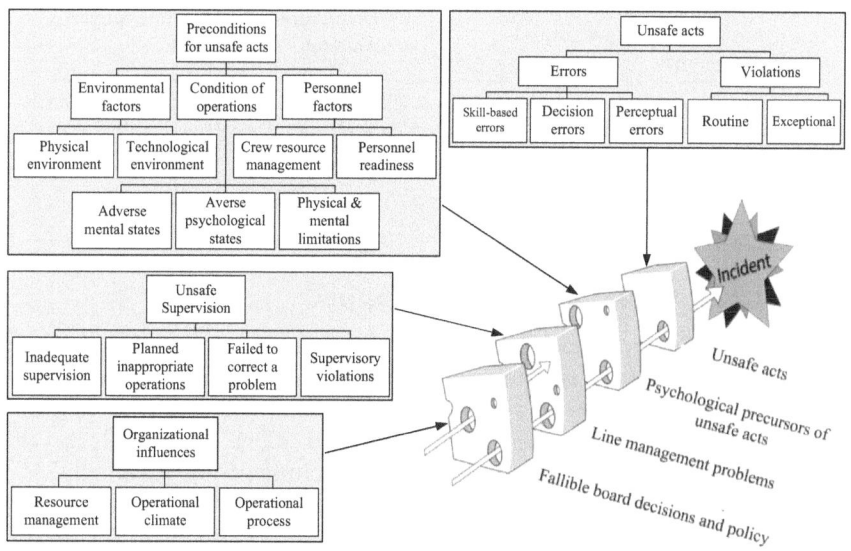

FIGURE 2.4
HFACS overlaid on Reason's Swiss Cheese model. (From Salmon, P. M., Cornelissen, M., & Trotter, M. J. 2012. Systems-based accident analysis methods: A comparison of Accimap, HFACS, and STAMP. *Safety Science, 50*(4), 1158–1170. doi:10.1016/j.ssci.2011.11.009. With permission.)

TABLE 2.1

Unsafe Acts Level External Error Mode Categories

Errors	Violations
Skill-Based Errors	**Routine**
• Breakdown in visual scan	• Inadequate briefing for flight
• Inadvertent use of flight controls	• Failed to use ATC radar advisories
• Poor technique/airmanship	• Flew an unauthorised approach
• Over-controlled the aircraft	• Violated training rules
• Omitted checklist item	• Filed VFR in marginal weather
• Omitted step in procedure	conditions
• Over reliance on automation	• Failed to comply with departmental
• Failed to prioritise attention	manuals
• Task overload	• Violation of orders, regulations, SOPs
• Negative habit	• Failed to inspect aircraft after in-flight
• Failure to see and avoid	caution light
• Distraction	**Exceptional**
	• Performed unauthorised acrobatic
Decision Errors	manoeuvre
• Inappropriate manoeuvre/procedure	• Improper takeoff technique
• Inadequate knowledge of systems &	• Failed to obtain valid weather brief
procedures	• Exceeded limits of aircraft
• Exceeded ability	• Failed to complete performance
• Wrong response to emergency	computations
	• Accepted unnecessary hazard
Perceptual Errors	• Not current/qualified for flight
• Due to visual illusion	• Unauthorised low altitude canyon
• Due to spatial disorientation/vertigo	running
• Due to misjudged distance, altitude,	
airspeed, clearance	

Source: From Wiegmann, D. A., & Shappell, S. A. 2003. *A human error approach to aviation accident analysis: The human factors analysis and classification system.* Burlington, VT: Ashgate.

2.5 Accimap, CAST, and HFACS Applied

To compare and contrast each method, we present an overview of the outputs produced by each when used to analyze an incident. The focus of the analysis is the Mangatepopo Gorge walking incident in which six students and their teacher drowned (Brookes, Smith, & Corkill, 2009). Three analysts with significant experience in the application of Accimap, STAMP, and HFACS independently used one of the methods to analyze the Mangatepopo Gorge Incident. Each analysis was based on the same data sources, including the independent investigation report (Brookes et al., 2009) and coroner's report (Davenport, 2010). Upon completion of the initial analyses, the three analysts exchanged and reviewed each other's analyses, with any discrepancies or disagreements' being resolved through discussion until consensus was reached.

To ensure external validity of the analyses produced, a subject matter expert reviewed the final analysis outputs.

To provide context for the outputs from the analyses, the incident is described in detail below. The outputs from the three methods are then presented.

2.5.1 Description of the Incident

The incident occurred on the 15th April 2008. A group of college students, their teacher, and an activity leader from an Outdoor Pursuit Centre (OPC) were canyoning in the Mangatepopo Gorge in Tongariro National Park, New Zealand. Following a period of heavy rain, a flash flood led to the group becoming trapped on a small ledge above the river. With the river level rising rapidly, the activity leader conceived a plan to evacuate the group from the gorge. This involved tying strong swimmers to weak swimmers (in pairs) and subsequently asking each pair to follow the activity leader into the river, in 5-minute intervals. As they were swept downstream, the activity leader planned to extract each pair using a 'throwbag' technique whereby a bag is attached to a length of rope, thrown to the person in the water, and used to pull them to the safety of the riverbank. After attempting the evacuation, only the activity leader and two students managed to get out of the river as intended, with the remaining eight students and their teacher being swept downstream. Tragically, six students and their teacher drowned after being swept over a spillway.

In the aftermath of the incident, the coroner and an independent investigation initiated by the OPC identified various contributory factors relating to the activity leader, the field manager, the activity environment, the OPC, the local weather service, and government legislation and regulation (e.g. Brookes et al., 2009; Davenport, 2010).

Immediate contributory factors included the increased river flow, and the activity leader's decision to remove the group from the ledge and enter the river. The activity leader's decision to leave the ledge was, of course, influenced by several factors. She was inexperienced, both generally as an activity leader, and specifically in the gorge. She was also in charge of a group that included several non-confident swimmers. The water level was visibly rising, the group was cold and uncomfortable, and some were questioning whether they should stay on the ledge (Brookes et al., 2009). The activity leader had no way of contacting the OPC's field manager, as the one radio in her possession did not work on the ledge due to poor signal reception.

Many other contributory factors were identified in the independent investigation and coroner's reports. The evacuation plan and instructions given to the group were deemed to be inadequate (Brookes et al., 2009). The connection of strong swimmers to weak swimmers was judged to be inappropriate, and did not consider that it would make the participants more vulnerable to drowning. The use of the throwbag technique was inappropriate given the adverse conditions (Brookes et al., 2009). No warning regarding the spill-way was given by the activity leader, and no advice was provided about what to do in the event of being washed over the spillway (Brookes et al., 2009). Consequently, the independent investigation report concluded that the group did not appreciate the gravity of the situation.

The first student entered the water only 1 to 2 minutes after the activity leader (not 5 minutes later as requested), which meant that the she was not ready to execute the throwbag rescue. The first student was subsequently swept over the spillway. Although the activity leader quickly made a radio emergency call to the OPC, she had no way of contacting the group on the ledge to tell them to stop and stay put.

Various factors help to explain why the activity leader's actions made sense to her at the time. She had little experience of leading gorge walk-ing activities and had no experience of the gorge in adverse weather con-ditions. As mentioned, she was unable to contact the field manager at the OPC for advice. The radio was not waterproof and had to be carried disassembled, in a protective bag. It was not assembled until the group were trapped on the ledge, and poor reception meant that it did not work well in the gorge.

According to the independent investigation report, the OPC's field man-ager was preoccupied with an audit that was taking place on the day of the incident (Brookes et al., 2009). During the morning staff meeting, the field manager did not check the maps on the MetService weather fax, which meant that he, along with the other OPC staff, were unaware of the impending bad weather. The weather fax used during the meeting was also incomplete, with the word 'thunderstorms' missing from the sentence 'Today rain with iso-lated and poor visibility at times' (Brookes et al., 2009). This led the field man-ager and activity leader to assume that the rain would ease in the afternoon (Brookes et al., 2009). An updated forecast, describing likely heavy rainfall and isolated thunderstorms, was available on the MetService website after 7.15am (Davenport, 2010); however, this was not checked following the morn-ing meeting by the field manager, activity leader, or other OPC staff.

As a result, the field manager did not cancel all gorge activities, nor did he state that downstream gorge trips should be closed (which he had decided based upon the weather forecast). Having raised concerns with the activity leader regarding the planned gorge walking activity, it appears that the field manager misunderstood the exact nature of the planned trip, believing that the activity leader planned to not go too far into the gorge (no more than 100 metres).

Moving higher into the OPC organization, the activity leader did not sign off for the activity on the risk assessment and management system, which provided information on environmental hazards. In addition, prior to the activity, the OPC had not determined the swimming capabilities of the stu-dents within the group. This was outlined in the OPC's policies as a require-ment before undertaking water activities (Brookes et al., 2009).

Other factors related to the OPC's policies, procedures, and programs were also identified. The OPC's induction, mentoring, and training programs were judged to be inadequate (Brookes et al., 2009). Further, the independent investigation report stated that the risk assessment system for the upstream gorge activity had no map of the area, did not identify the spillway as a hazard, and provided no information on previous incidents (Brookes et al., 2009). In addition, there was no rescue procedure for incidents in the gorge, despite the fact that incidents had occurred there previously. Finally, the OPC's incident reporting and learning system was criticised. Although various incidents had occurred in the gorge previously, and there was a record of them, the information was not formally communicated to staff, or analyzed to identify trends or safety issues.

Further factors also influenced the OPC's operations. They were under significant financial and production pressures, which ostensibly led to issues with the design of their adventure programs, pressure to get staff trained and competent for activities, and the use of only one activity leader for activities during busy periods (Brookes et al., 2009). Further, as with most activity centres, the OPC had a high staff turnover, which in turn meant organizational memory relating to incidents, and risk in the gorge, was not maintained. The centre's 'rain or shine' culture was also highlighted in the independent investigation report as a factor that influenced operations and staff behavior.

Finally, factors outside of the OPC were also identified. There was no regulatory or licensing body for outdoor activity centres. The auditing system came under scrutiny, as the auditor failed to intervene at any point during the day in question. Further, none of the issues discussed above were mentioned in the auditor's report (Brookes et al., 2009). There was also no legislation covering outdoor activity centres and led outdoor activities.

2.5.2 Accimap Analysis of the Mangatepopo Gorge Incident

The Accimap analysis of the Mangatepopo Gorge Incident is presented in Figure 2.5. The Accimap was able to incorporate all of the contributory factors identified in the independent investigation and coroner's reports, and also shows the relationships between contributory factors within and across all levels of the system.

2.5.3 CAST Analysis of the Mangatepopo Gorge Incident

The control structure for the incident is presented in Figure 2.6. This control structure included the auditor, the OPC company, the OPC chief executive, the OPC manager, the field manager, the training manager, instructors, students and teachers, and the students' parents.

Examples of the contributory factors related to field manager and activity leader control and feedback mechanisms are presented in Figures 2.7 and 2.8,

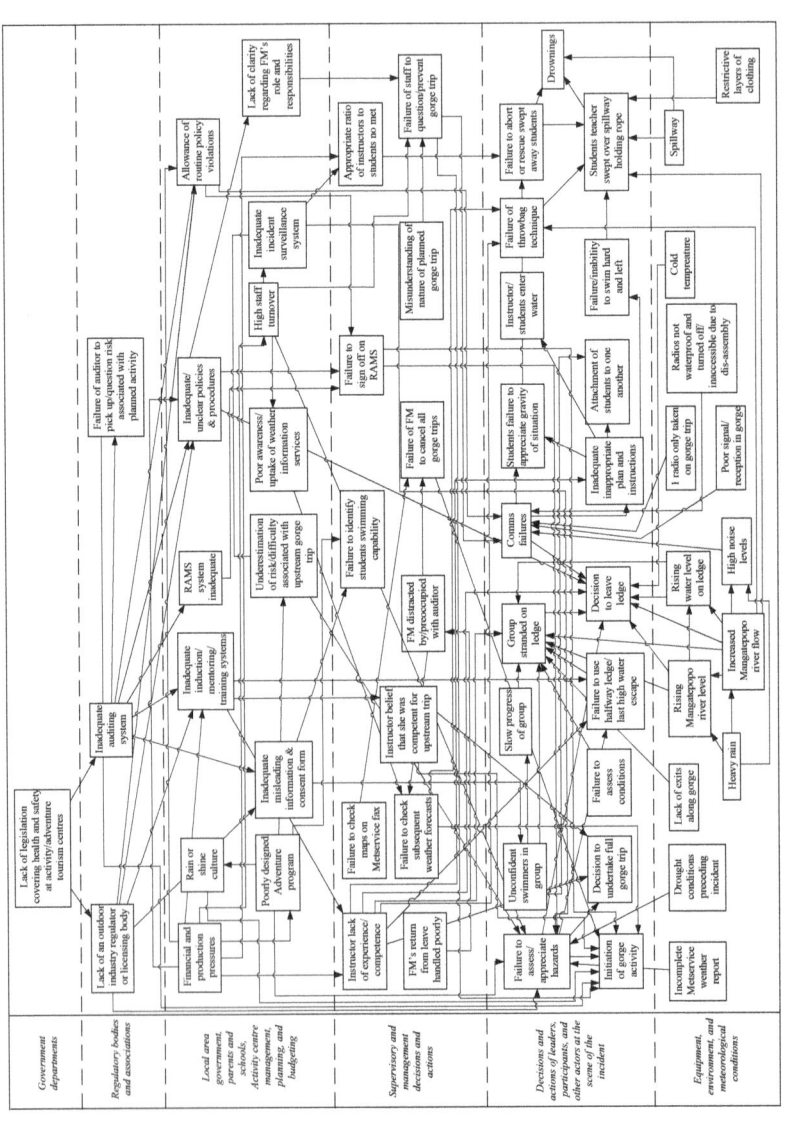

FIGURE 2.5
Accimap of the Mangatepopo Gorge Incident. (From Salmon, P. M., Cornelissen, M., & Trotter, M. J. 2012. Systems-based accident analysis methods: A comparison of Accimap, HFACS, and STAMP. *Safety Science, 50*(4), 1158–1170. doi:10.1016/j.ssci.2011.11.009. With permission.)

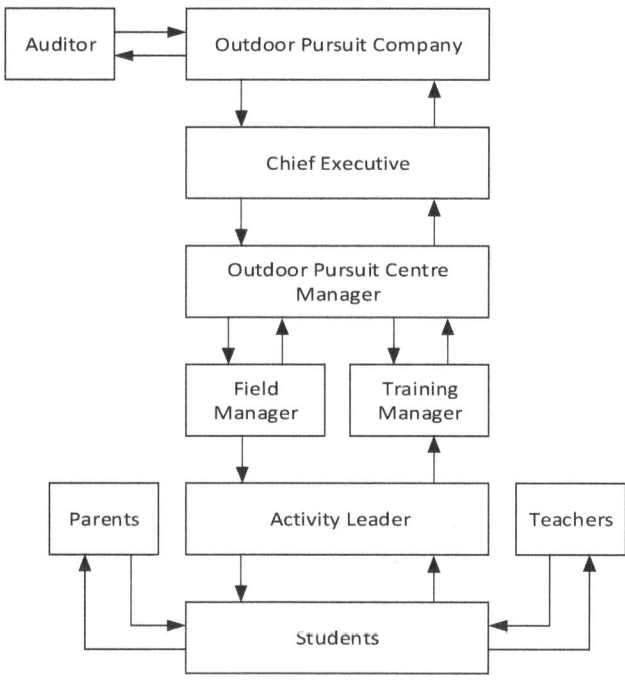

FIGURE 2.6
CAST analysis of Mangatepopo Gorge Incident control structure diagram.

respectively. Whilst CAST was able to support classification of contributory factors related to the different actors and organizations within the control structure, it was often difficult to distinguish between inadequate control actions and inadequate enforcement of constraints when making this classification. In addition, contributory factors were often repeated across the different control, feedback and mental model failure categories. For example, in Figure 2.8 the contributory factor relating to the fact that the activity participants' swimming ability was not assessed is presented both as an 'inadequate enforcement of constraints' and an 'inadequate control action', whilst at the same time being expressed in the 'inadequate or missing feedback' and 'mental model flaw' categories. Again, this relates to the difficulty for analysts to distinguish between the CAST failure categories (Figure 2.3).

A key omission from the analysis is a representation of the relationships between contributory factors. A final limitation to note is that the control structure produced lacks detail in terms of the specific control and feedback mechanisms that were in place. This relates more to the data available for the analysis, however, it is worth noting that comprehensive control structure analyses require significant resources to develop in terms of data, access to subject matter experts, and time.

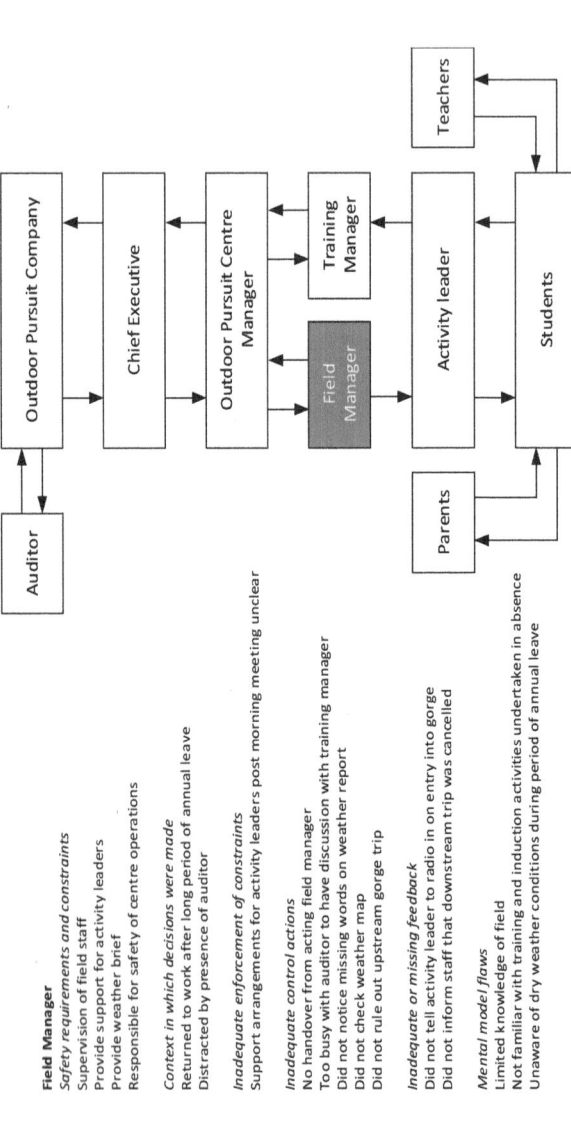

Field Manager

Safety requirements and constraints
Supervision of field staff
Provide support for activity leaders
Provide weather brief
Responsible for safety of centre operations

Context in which decisions were made
Returned to work after long period of annual leave
Distracted by presence of auditor

Inadequate enforcement of constraints
Support arrangements for activity leaders post morning meeting unclear

Inadequate control actions
No handover from acting field manager
Too busy with auditor to have discussion with training manager
Did not notice missing words on weather report
Did not check weather map
Did not rule out upstream gorge trip

Inadequate or missing feedback
Did not tell activity leader to radio in on entry into gorge
Did not inform staff that downstream trip was cancelled

Mental model flaws
Limited knowledge of field
Not familiar with training and induction activities undertaken in absence
Unaware of dry weather conditions during period of annual leave

FIGURE 2.7
CAST analysis of the Mangatepopo Gorge Incident. Examples of contributory factors related to field manager–related control and feedback mechanisms.

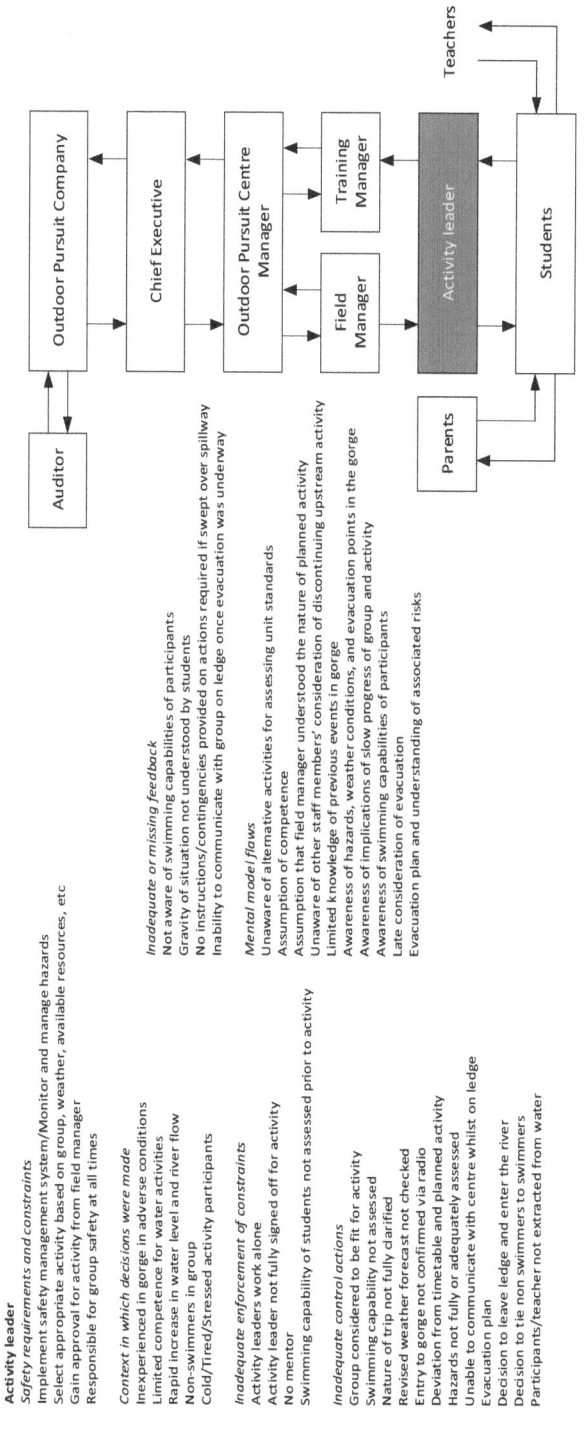

Activity leader

Safety requirements and constraints
Implement safety management system/Monitor and manage hazards
Select appropriate activity based on group, weather, available resources, etc
Gain approval for activity from field manager
Responsible for group safety at all times

Context in which decisions were made
Inexperienced in gorge in adverse conditions
Limited competence for water activities
Rapid increase in water level and river flow
Non-swimmers in group
Cold/Tired/Stressed activity participants

Inadequate enforcement of constraints
Activity leaders work alone
Activity leader not fully signed off for activity
No mentor
Swimming capability of students not assessed prior to activity

Inadequate control actions
Group considered to be fit for activity
Swimming capability not assessed
Nature of trip not fully clarified
Revised weather forecast not checked
Entry to gorge not confirmed via radio
Deviation from timetable and planned activity
Hazards not fully or adequately assessed
Unable to communicate with centre whilst on ledge
Evacuation plan
Decision to leave ledge and enter the river
Decision to tie non swimmers to swimmers
Participants/teacher not extracted from water

Inadequate or missing feedback
Not aware of swimming capabilities of participants
Gravity of situation not understood by students
No instructions/contingencies provided on actions required if swept over spillway
Inability to communicate with group on ledge once evacuation was underway

Mental model/flows
Unaware of alternative activities for assessing unit standards
Assumption of competence
Assumption that field manager understood the nature of planned activity
Unaware of other staff members' consideration of discontinuing upstream activity
Limited knowledge of previous events in gorge
Awareness of hazards, weather conditions, and evacuation points in the gorge
Awareness of implications of slow progress of group and activity
Awareness of swimming capabilities of participants
Late consideration of evacuation
Evacuation plan and understanding of associated risks

FIGURE 2.8

CAST analysis of the Mangatepopo Gorge Incident. Examples of contributory factors related to activity leader-related control and feedback mechanisms.

2.5.4 HFACS Analysis of the Mangatepopo Gorge Incident

The HFACS analysis is presented in Figure 2.9. Unlike both Accimap and CAST, the HFACS analysis was only able to identify contributory factors within the organization, and so did not consider factors related to the auditing system, or regulation and legislation surrounding the provision of outdoor activities. In addition, the HFACS analysis was unable to classify and represent relationships between contributory factors. A final notable feature of the HFACS analysis is the unavoidable use of negative terminology as a result of the focus of the classification scheme on errors, failures, and violations (e.g. failure, inadequate, poor).

2.5.5 Comparison of Findings

In summary, comparison of the three analyses highlights important differences between the methods in terms of both the approach taken and the outputs produced. These are drawn out below by discussing each method in relation to the systems thinking principles for incident reporting outlined at the end of Chapter 1.

2.5.5.1 Principle 1: 'Up and Out' Rather Than 'Down and In'

As discussed in Chapter 1, to be effective, incident reporting systems need to collect data on contributory factors from across the overall work system. Consequently, a method used as part of an incident reporting system should be able to identify and describe contributory factors across the overall system. Both Accimap and CAST explicitly encourage the identification of contributory factors across the overall work system by incorporating a focus on levels of the system, up to, and including, government. HFACS enables analysts to identify contributory factors up to the organizational level; however, the original method does not consider contributory factors outside of the organizations involved. In the present case study for example, it was not possible to classify contributory factors relating to the auditor who was present at the OPC on the day of the incident. More recent versions of HFACS have used additional 'outside factors' or 'external influences' levels to include regulatory factors, legislation gaps, design flaws, and administration oversights (Chen et al., 2013; Patterson & Shappell, 2010). In addition, HFACS encourages a strong focus on errors made by human operators at the sharp end of the system by providing a detailed 'unsafe acts' taxonomy that typically only applied to frontline operators (e.g. pilots, drivers). In the present case study, for example, skill-based errors, decision errors and perceptual errors associated with the activity leader were identified using HFACS.

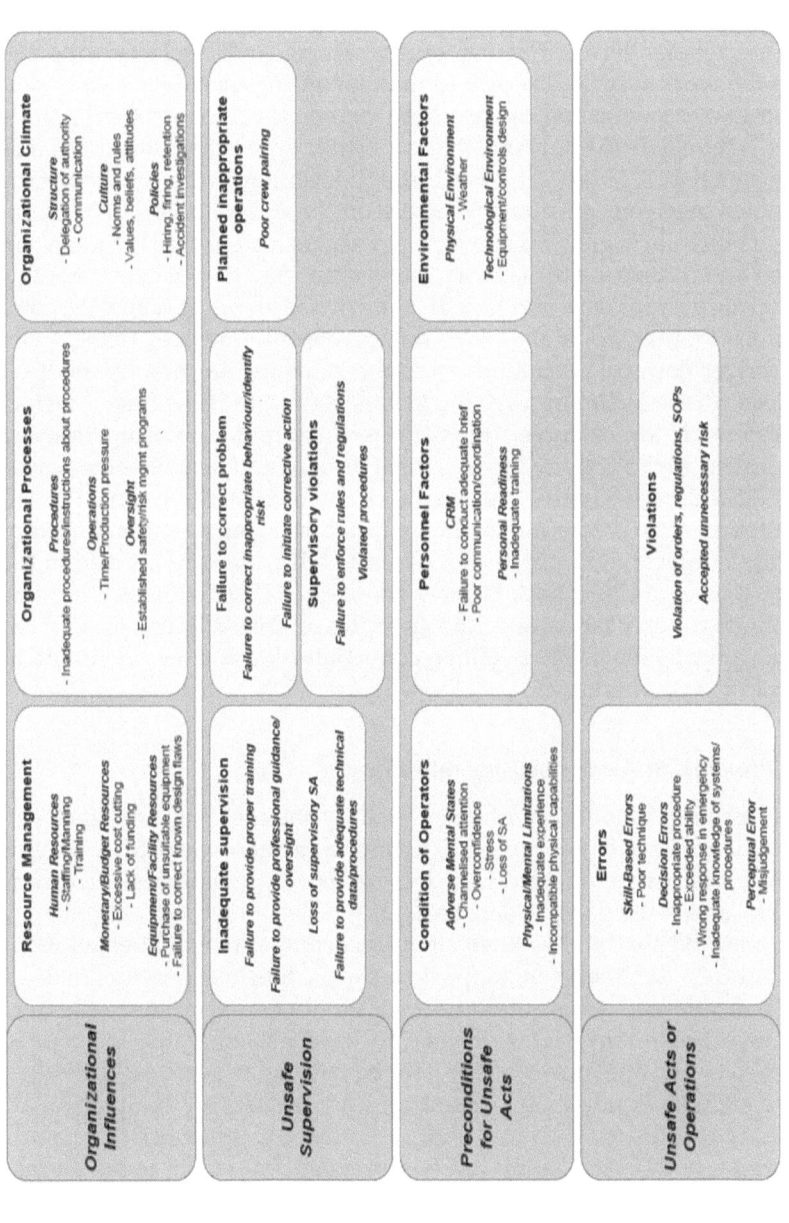

FIGURE 2.9
HFACS analysis of the Mangatepopo Gorge Incident.

2.5.5.2 *Principle 2: Interactions and Relationships*

A key element of systems thinking is the notion that the interactions between components are more important than the behavior of components themselves. This relates to the idea that incidents are emergent properties created by the interactions between actors, organizations, and artefacts across all levels of the work system. The key to understanding incidents is to understand these interactions and the resulting emergent behaviors. Of the three methods, Accimap is the only one that explicitly represents interactions and emergent behaviors. This is achieved through enabling analysts to depict the relationships between contributory factors on the Accimap itself. For example, the Accimap in Figure 2.5 shows that various factors across the system interacted and influenced the activity leader's decision to evacuate the ledge, including her inexperience and lack of competence for gorge activities, communications failures (i.e. inability to contact centre for advice), rising water level and river flow, cold conditions, and the students questioning the decision to stay on the ledge. In turn, the factors related to these linked factors are represented; for example, the various induction/mentoring/training failures linked to the activity leader's inexperience and lack of competence. Whilst CAST depicts control and feedback interactions between different levels of the system, it does not explicitly link actors, organizations, or artefacts to these interactions, nor does it describe what interactions lead to the emergent control and feedback failures. Similarly, HFACS does not provide a mechanism for analysts to specifically represent interactions between the contributory factors identified. Rather, contributory factors are identified in isolation from one another.

2.5.5.3 *Principle 3: Avoid Focusing on Failure*

Systems thinking models are underpinned by the idea that incidents can happen when no parts are broken. Therefore, incident reporting systems should be blame free, and avoid focusing exclusively on failures (e.g. errors, violations, lapses), or gaps in defences. Of the three methods, Accimap is the only method that allows analysts to incorporate normal behaviors as well as failures. Accimap can support this as it does not provide analysts with classification schemes of failure modes; rather, analysts have the freedom to incorporate any factor deemed to have played a role incident in question. Accimap analyses can therefore be tailored to avoid failure-based language such as 'inadequate', 'failed to', and 'failures in' (Salmon et al., 2015).

In contrast, both CAST and HFACS encourage an explicit focus on failures. With CAST, the analyst uses a taxonomy of control and feedback failures to classify problems with the system's control and feedback mechanisms. Likewise, HFACS uses taxonomies of error and failure modes across four system levels: unsafe acts, preconditions for unsafe acts, unsafe supervision, and organizational influences. The use of error or failure-based taxonomies by both methods means that there is little scope for analysts to include behaviors other than those deemed to have been failures of some sort. There is no opportunity for analysts to incorporate normal behaviors in their descriptions of incidents – they must force fit events into one of the error or failure modes provided. This is inappropriate given current knowledge on incident causation, especially as a worrying consequence may be that the normal behaviors that contribute to incident are not picked up during analyses. This could impact incident prevention activities by providing a false sense of security that nothing else is involved, and nothing needs fixing (apart from error-producing human operators).

2.5.5.4 *Principle 4: A Systems Lens*

There is a need to provide practitioners with an appropriate methodological framework with which to report, analyze, and represent incidents from a systems perspective. Whilst all three methods are generally underpinned by systems thinking, HFACS arguably does not incorporate many of the accepted elements of a systems approach to incident causation. As discussed, there is a focus on unsafe acts, errors, and violations, and the method does not enable any interactions between components to be considered. Accimap and STAMP, on the other hand, are underpinned by Rasmussen's risk management framework, which is arguably the dominant systems thinking model of accident causation. In the present case study, this enabled the Accimap and CAST analysts to consider the overall system, including actors and organizations outside of the OPC.

2.5.6 A Note on Reliability

In addition to the systems thinking characteristics described above, an analysis method must also be reliable to prove useful in incident analysis and prevention efforts. A method is reliable if it produces the same outcome when used on different occasions by the same analysts and by different analysts (i.e. intra- and inter-analyst reliability; Stanton, 2016; Stanton & Young, 1999, 2003). Although the CAST and HFACS classification schemes introduce problems, they are useful for strengthening the reliability of analysis methods. Reliability is enhanced since analysts are given more guidance about possible types of contributory factors, and the boundary around the analysis. The positive impact of classification schemes on reliability is evidenced by various HFACS studies that report statistics demonstrating acceptable levels of inter-rater reliability (e.g. Lenné, Salmon, Liu, & Trotter, 2012; Li & Harris, 2006; Li, Harris, & Yu, 2008). The use of a classification scheme also supports a methods use in analyses of multiple incidents (as would be required in an incident reporting system), since themes and trends across incidents can be easily determined by analysing the frequency of contributory factors. This is evidenced by the frequent use of HFACS in multiple incident analyses (e.g. Baysari, McIntosh, & Wilson, 2008; ElBardissi, Wiegmann, Dearani, Daly, & Sundt, 2007; Li & Harris, 2006; Li et al., 2008; Patterson & Shappell, 2010), whilst most published Accimap analyses focus on single incidents (e.g. Cassano-Piche et al., 2009; Jenkins, Salmon, Stanton, & Walker, 2010; Johnson & de Almeida, 2008; Salmon, Williamson, Lenné, Mitsopoulos-Rubens, & Rudin-Brown, 2010).

The absence of a classification scheme to describe the contributory factors at each level of Accimap is therefore problematic. The analysis is entirely dependent upon an analyst's subjective judgement, and the reliability of the method is likely to be limited. Differences in both the way actual contributory factors are identified, and the way in which the contributory factors are described, are likely to emerge across different analyses. Without a classification scheme, it is difficult to aggregate Accimap analyses to derive a useful summary of multiple incidents. It should be noted though, that a poorly developed or limited classification scheme will cause more problems than it solves, as analysts will be limited in their analyses of incidents. The careful development of a comprehensive, and rigorous, classification scheme is therefore required to enhance Accimap's suitability for use as part of an incident reporting system (Salmon et al., 2012).

2.6 Summary and Conclusions

The aim of this chapter was to compare and contrast three currently popular systems thinking–based analysis methods: Accimap, HFACS, and STAMP. The intention was to identify which of the three approaches is the most suited for use as part of an incident reporting system. Based on an analysis of the Mangatepopo Gorge Incident, Accimap was identified as the most useful analysis method, and best suited for use as part of incident reporting systems. Out of the three methods, Accimap best achieved the key incident reporting system requirements outlined in Chapter 1, and produced the richest and most useful description of the Mangatepopo Gorge Incident (Salmon et al., 2012). Based on the need to enhance reliability, it was recommended that incident reporting systems using Accimap should incorporate a suitable classification scheme of contributory factors to support analysts in identifying and classifying contributory factors, as well as aggregating analyses. The development of such a classification scheme for the led outdoor activity domain is described in Chapters 6 and 7.

References

Baysari, M. T., McIntosh, A. S., & Wilson, J. R. (2008). Understanding the human factors contribution to railway accidents and incidents in Australia. *Accident Analysis & Prevention, 40*(5), 1750–1757. doi:10.1016/j.aap.2008.06.013

Brookes, A., Smith, M., & Corkill, B. (2009). Report to the trustees of the Sir Edmund Hillary Outdoor Pursuit Centre of New Zealand, Mangatepopo Gorge Incident, 15 April 2008. *Turangi: OPC Trust.*

Cassano-Piche, A. L., Vicente, K. J., & Jamieson, G. A. (2009). A test of Rasmussen's risk management framework in the food safety domain: BSE in the UK. *Theoretical Issues in Ergonomics Science, 10*(4), 283–304. doi:10.1080/14639220802059232

Chen, S.-T., Wall, A., Davies, P., Yang, Z., Wang, J., & Chou, Y.-H. (2013). A Human and Organizational Factors (HOFs) analysis method for marine casualties using HFACS-Maritime Accidents (HFACS-MA). *Safety Science, 60*, 105–114. doi:10.1016/j.ssci.2013.06.009

Davenport, C. J. (2010). *Mangatepopo Coroners report.* Retrieved from http://outdoor council.asn.au/doc/Coroners_Report_OPC.pdf. Access date: 30th March 2010.

ElBardissi, A. W., Wiegmann, D. A., Dearani, J. A., Daly, R. C., & Sundt, T. M. (2007). Application of the human factors analysis and classification system methodology to the cardiovascular surgery operating room. *Annals of Thoracic Surgery, 83*(4), 1412 1419. doi:10.1016/j.athoracsur.2006.11.002

Jenkins, D. P., Salmon, P. M., Stanton, N. A., & Walker, G. H. (2010). A systemic approach to accident analysis: A case study of the Stockwell shooting. *Ergonomics, 53*(1), 1–17. doi:10.1080/00140130903311625

Johnson, C. W., & de Almeida, I. M. (2008). An investigation into the loss of the Brazilian space programme's launch vehicle VLS-1 V03. *Safety Science, 46*(1), 38–53. doi:10.1016/j.ssci.2006.05.007

Lenné, M. G., Salmon, P. M., Liu, C. C., & Trotter, M. (2012). A systems approach to accident causation in mining: An application of the HFACS method. *Accident Analysis & Prevention, 48*, 111–117. doi:10.1016/j.aap.2011.05.026

Leveson, N. G. (2004). A new accident model for engineering safer systems. *Safety Science, 42*(4), 237–270. doi:10.1016/S0925-7535(03)00047-X

Li, W.-C., & Harris, D. (2006). Pilot error and its relationship with higher organizational levels: HFACS analysis of 523 accidents. *Aviation, Space, and Environmental Medicine, 77*(10), 1056–1061.

Li, W.-C., Harris, D., & Yu, C.-S. (2008). Routes to failure: Analysis of 41 civil aviation accidents from the Republic of China using the human factors analysis and classification system. *Accident Analysis & Prevention, 40*(2), 426–434. doi:10.1016/j .aap.2007.07.011

Patterson, J. M., & Shappell, S. A. (2010). Operator error and system deficiencies: Analysis of 508 mining incidents and accidents from Queensland, Australia using HFACS. *Accident Analysis & Prevention, 42*(4), 1379–1385. doi:10.1016/j.aap.2010.02.018

Rasmussen, J. (1997). Risk management in a dynamic society: A modelling problem. *Safety Science, 27*(2–3), 183–213. doi:10.1016/S0925-7535(97)00052-0

Salmon, P. M., Cornelissen, M., & Trotter, M. J. (2012). Systems-based accident analysis methods: A comparison of Accimap, HFACS, and STAMP. *Safety Science, 50*(4), 1158–1170. doi:10.1016/j.ssci.2011.11.009

Salmon, P. M., Goode, N., Stevens, E., Walker, G. H., & Stanton, N. A. (2015, August 2–7, 2015). *The elephant in the room: Normal performance and accident analysis.* Paper presented at the International Conference on Engineering Psychology and Cognitive Ergonomics, Los Angeles, CA, USA.

Salmon, P. M., Stanton, N. A., Lenné, M. G., Jenkins, D. P., Rafferty, L., & Walker, G. H. (2011). *Human factors methods and accident analysis: Practical guidance and case study applications.* Aldershot, UK: Ashgate.

Salmon, P. M., Williamson, A., Lenné, M., Mitsopoulos-Rubens, E., & Rudin-Brown, C. M. (2010). Systems-based accident analysis in the led outdoor activity domain: Application and evaluation of a risk management framework. *Ergonomics, 53*(8), 927–939. doi:10.1080/00140139.2010.489966

Stanton, N. A. (2016). On the reliability and validity of, and training in, ergonomics methods: A challenge revisited. *Theoretical Issues in Ergonomics Science, 17*(4), 345–353. doi:10.1080/1463922X.2015.1117688

Stanton, N. A., & Young, M. S. (1999). What price ergonomics? *Nature, 399*(6733), 197. doi:10.1038/20298

Stanton, N. A., & Young, M. S. (2003). Giving ergonomics away? The application of ergonomics methods by novices. *Applied Ergonomics, 34*(5), 479–490. doi:10.1016/S0003-6870(03)00067-X

Svedung, I., & Rasmussen, J. (2002). Graphic representation of accident scenarios: Mapping system structure and the causation of accidents. *Safety Science.*

Wiegmann, D. A., & Shappell, S. A. (2003). *A human error approach to aviation accident analysis: The human factors analysis and classification system.* Burlington, VT: Ashgate.

3

A Process Model for Developing an Incident Reporting System

Practitioner Summary

This chapter presents a process model for developing an incident reporting system that is practical for end users, produces good quality data, and underpinned by the principles of systems thinking. The process involves six key stages: (1) understanding the context of use; (2) developing a domain-specific accident analysis method; (3) designing a data collection protocol; (4) designing a process for collecting, analyzing and translating incident reports into actions; (5) developing supporting software tools and training materials; and (6) evaluating the implementation of the system. The first five stages involve the development of specific components, which together form a comprehensive system for collecting, analyzing, and learning from incident reports. The chapter concludes by summarizing the optimal characteristics of each component in the process model. These characteristics can be used to guide the design of new incident reporting systems developed using the process described in this chapter, or as criteria for evaluating the strengths and weaknesses of existing incident reporting systems.

3.1 Introduction

Chapter 1 outlined four core principles of the systems thinking approach for the design of incident reporting systems:

- *Principle 1: 'Look up and out' rather than 'down and in'* by collecting data on contributory factors from across the work system.

- *Principle 2: Identify interactions and relationships* by encouraging users to provide data regarding potential interactions between contributory factors both within, and across, system levels.

- *Principle 3: Avoid focusing exclusively on failures* by enabling users to describe how the normal conditions of work contributed to incidents, as well as identifying perceived problems, or gaps in defences.
- *Principle 4: Apply a systems lens* by including a methodological framework to report, analyze, and represent incidents from a systems thinking perspective.

This chapter presents a process model for developing an incident reporting system underpinned by these principles, which also collects good quality data, and is practical for end users.

3.2 Overview of the Process Model

A overview of the process model is presented in Figure 3.1. The model describes the core components of an incident reporting system underpinned by systems thinking, and a process for refining them through iterative cycles of end-user testing. The process model is applicable to the development of both internal incident reporting systems, collecting data within a single organization, and sector-wide incident reporting systems, collecting data from multiple organizations.

The following section describes the rationale behind the development sequence. The stages of the development process are then described, with specific guidance on the optimal design of each component in the process model. This guidance is based on the systems thinking principles, good practice guidelines for injury surveillance and incident reporting systems, and the literature on supporting learning from incidents within organizations (Goode, Finch, Cassell, Lenné, & Salmon, 2014; Goode, Salmon, Newnam, Dekker, Stevens, & van Mulken, 2016).

3.3 The Development Sequence

The development sequence shown in Figure 3.1 is a complete reversal of how incident reporting systems have historically been developed. In the majority of incident reporting systems, the incident report form (i.e. data collection fields) was developed first. The classification scheme for analyzing contributory factors, and aggregating the findings, was then only developed after the sheer quantity of incident reports became overwhelming (Johnson, 2002; Wallace & Ross, 2006b). This traditional development sequence makes little sense because the types of contributory factors that are identified from incident reports are highly dependent on the data collection fields (Salmon et al., 2011;

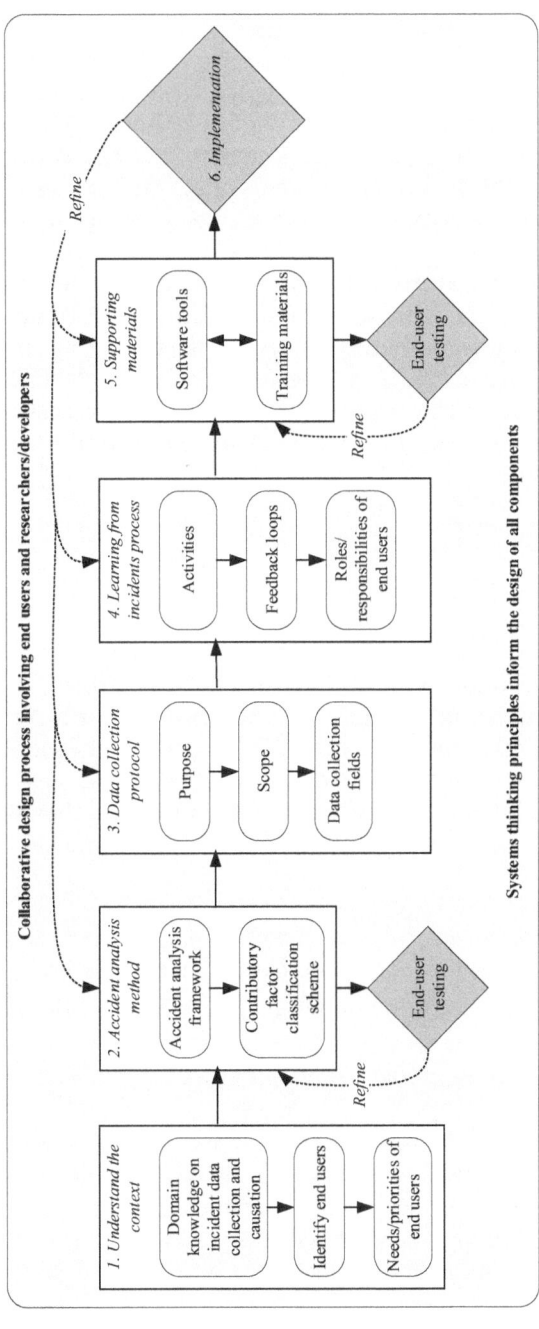

FIGURE 3.1

Process model for developing an incident reporting system underpinned by the principles of system thinking.

Underwood & Waterson, 2013). It is highly unlikely that an incident report form developed on an ad hoc basis will collect sufficiently detailed, appropriate data for a systems thinking analysis.

To feed a systems thinking analysis, the data collection fields on the incident report form must be explicitly designed to support the identification of relevant details (Salmon et al., 2011; Wallace & Ross, 2006b). As reflected in the process model (Figure 3.1), this means that the accident analysis method must be developed first. The data collection protocol and software tools should then be developed to meet the needs of the accident analysis method.

Another important feature of the process model is the involvement of end users in all stages of the development. The design of all the components needs to be underpinned by a good understanding of the context, and end-user requirements. End users should be involved in the initial design of each component, and formal assessment processes should be used to ensure they are practical for end users, and support the collection of good quality data.

3.4 Stage 1: Understand the Context

Developing a good understanding of the context is critical. Although this might seem obvious, prior to beginning the development of a new incident reporting system, it is helpful to meet with end users in the place where they will use the new system, and observe them using existing systems. This will provide early insights into workflows, and the problems they might face in using the new system. Three formal tasks need to be completed during this stage:

1. Review *existing knowledge* on incident data collection and causation in the relevant domain (e.g. healthcare, construction, etc.);
2. Develop an understanding of the *end users* of the incident reporting system; and
3. Identify the *needs* and *priorities* of end users.

Reviewing existing knowledge is required to clearly establish the need for the new incident reporting system. The review should determine whether there are existing incident reporting systems, accident analysis methods, or contributory factor classification schemes that align with systems thinking, and so could be utilized, or at least inform the development of the new system. Chapter 4 provides further guidance on reviewing and evaluating existing knowledge in a domain.

Once the need for developing a new incident reporting system is established, the next task is developing an understanding of the *end users* of the proposed system. This is important because (a) they need to

ACCEPTABILITY

The system matches the characteristics, needs, and priorities of the end users.

be involved in the development process; and (b) this understanding needs to inform the design of all of the components in the system. This is the only way to ensure that the end product is both *acceptable* and usable in practice.

A good starting point is describing end users based on their *roles, tasks,* and *goals* within the incident reporting system (Maguire, 2001). For any given incident reporting system, there are likely to be several different types of end users (i.e. roles) with different tasks and goals (Johnson, 2003). For example, an incident reporting system might be used by *staff* to report incidents; *managers* to view incident reports; *safety managers* to analyze the data; and the broader *organization* or *industry stakeholders* to identify incident prevention strategies. This information should then inform who should be involved in the development of each component of the system. For example, staff and managers may need to be involved in the design and testing of an incident report form, whereas safety managers are usually required for the design and testing of an accident analysis method.

It is also important to consider the *prior knowledge* and *experiences* that will inform end-user interactions with the new incident reporting system (Baber, 2002). As shown in Figure 3.2, there are four key dimensions which influence how end users interact with an incident reporting system: (1) knowledge of the domain; (2) experience with computers; (3) experience with other incident reporting systems; and (4) understanding of incident causation. These all need to be considered in the design of any new system.

End users' understanding of incident causation is particularly important for an incident reporting system underpinned by systems thinking. If end users think that human error is the primary of cause of incidents, then the outputs from the incident reporting system will likely conform to this view. Consequently, the design of the incident reporting system needs to ensure that end users understand that incidents are caused by interactions across the work system, including the normal conditions of work.

Finally, and most importantly, the needs and priorities of end users need to be clearly understood at the start of the development process. For example, some end users might place the most emphasis on minimizing workload, while others might be more concerned with maximizing data quality. This is critical information for the design of the components within an incident reporting system. Chapter 5 presents a process for formally identifying the needs and priorities of end users.

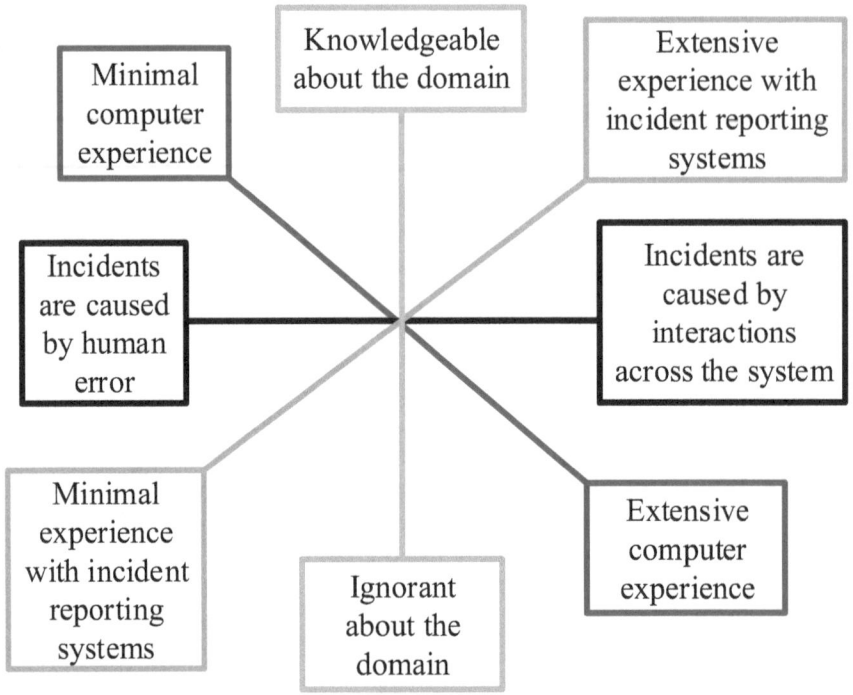

FIGURE 3.2
The four key dimensions on which end users' experience differs: (1) knowledge about computers in general; (2) experience with incident reporting systems (3) knowledge of the particular
domain; and (4) understanding of the systems thinking approach.

3.5 Stage 2: The Accident Analysis Method

The next stage is the selection, and adaptation, of an appropriate accident analysis method. As described in Chapter 2, Accimap (Svedung & Rasmussen, 2002)
is arguably the most suitable systems thinking–based accident analysis method
for use as part of an incident reporting system. The remainder of this book
therefore focuses on adapting Accimap for use in an incident reporting system.*

There are two tasks involved in
adapting Accimap to support the
reliable analysis of incidents in an
incident reporting system. The first
task involves adapting the labels on
the levels on the Accimap framework
to better describe the structure of the

USEFULNESS

The classification scheme includes
sufficiently detailed categories so that
aggregate analyses of incidents can
easily be interpreted.

---

* A similar process could be used to implement STAMP or HFACS in an incident reporting system, although their existing classification schemes would need to be adapted for the domain
of application.

system in question (e.g. healthcare, rail, manufacturing). Chapter 6 describes how to adapt the Accimap framework to describe a specific system.

The second task involves developing a domain-specific classification scheme describing the contributory factors found at each level of the Accimap framework. This is important as a classification scheme is required to support analysts in reliably classifying contributory factors across multiple incident reports (Salmon, Cornelissen, & Trotter, 2012).

The first step is developing a prototype classification scheme based on the available evidence in the domain, such as research on incident causation, investigation or inquiry reports, incident reports, multi-incident analyses, and existing classification schemes. During the development of the prototype classification scheme, the aim is to ensure that the categories are **useful** for detecting problems, and preventing future incidents. The categories need to be sufficiently detailed so that aggregate analyses of incidents are easily interpreted. A process for developing a prototype classification scheme is described in Chapter 6.

The prototype classification scheme then needs to be tested, and refined, with relevant end users (i.e. who will analyze the incident reports) until it produces:

- *Accurate* results (it needs to have satisfactory **validity**, i.e. end users classify contributory factors in line with expert opinion); and
- *Reproducible* results (it needs to be **reliable**, i.e. the same end user classifies contributory factors in the same way on repeated occasions, and different end users classify contributory factors in the same way).

> A classification scheme must show satisfactory **validity and reliability** before implementation in an incident reporting system.

A process for evaluating the reliability and validity of a contributory factor classification scheme is described in Chapter 7.

3.6 Stage 3: Data Collection Protocol

The third stage of the development process involves designing a data collection protocol, which is formal documentation describing:

1. The purpose of the incident reporting system;
2. The scope of the incident reporting system; and
3. The required data collection fields.

The following sections describe the important factors to consider in their design. Chapter 8 describes a process for designing a data collection protocol for an incident reporting system.

3.6.1 Purpose

The purpose of the system should be clearly defined to ensure that reporters understand how the data they contribute will be used (Leape & Abookire, 2005).

CLEAR PURPOSE AND OBJECTIVES

There is a clearly defined reason why the incident reporting system exists and how the data will be used.

While an incident reporting system may serve many purposes, the central purposes are typically related to either *accountability* or *learning*. Systems focused on *accountability* are established to monitor compliance with rules (e.g. procedures, regulation, legislation, etc.) and assign responsibility for incidents. For example, many countries require employers to report occupational injuries to identify non-compliance with occupational health and safety legislation (Lindberg, Hansson, & Rollenhagen, 2010).

In contrast, systems focused on *learning* are established to extract experiences from incidents, and convert them into measures and activities which will help to avoid future incidents (Jacobsson, Ek, & Akselsson, 2011). The primary purpose in a

NON-PUNITIVE

Reporters are free from the fear of retaliation or punishment as the result of reporting.

learning-orientated system is not to assign blame, but to inform system redesign (Leape & Abookire, 2005).

Although there are often attempts to utilize incident reporting systems for both purposes simultaneously, accountability and learning are not really compatible goals (Leape & Abookire, 2005). People are unwilling to report incidents if they fear retribution, punishment, or blame (Dekker, 2012; Hewitt, 2013; Sanne, 2008). It is therefore recommended that the purpose of incident reporting systems should be explicitly *non-punitive*, and focused on learning.

3.6.2 Scope

The type of incidents that should be reported must be clearly defined to optimize the quality of the data collected. Simply stating 'all incidents should be reported' is unlikely to be sufficient, as a significant barrier to

CLEAR CASE DEFINITIONS

There are clear definitions of what should and should not be reported.

reporting is often the belief that certain types of incidents or injuries are 'normal', and not worth reporting (Dekker, 2011; Vaughan, 1997).

Using the term 'error', in the definition of the scope, should also be avoided. From a psychological point of view, many errors (e.g. slips, lapses, and mistakes) are unintentional (Reason, 1997), and are therefore undetect-able, and unreportable if they do not result in obvious harm. Using the term 'error' also implies that an individual is to blame for the incident, which, again, may make people unwilling to report. Most importantly, one of the core principles of the systems thinking approach is that incidents are not necessarily the result of errors or failures – incidents may occur when the system is functioning as expected (Dekker, 2011).

Providing clear case definitions, with examples of the type of events that should, and should not, be reported helps provide clarity for reporters, and can increase the *sensitivity* and *specificity* of the data (See Section 3.9). For example, an occupational health and safety incident reporting system might specify that staff should report any pain, discomfort, or injuries associated with work tasks (e.g. shoulder pain attributed to computer use), or that occur during work hours (e.g. vehicle crashes during working hours). This type of system would exclude reports of pain, discomfort, or injury associated with activities conducted outside work hours (e.g. sporting events).

To optimize opportunities for learning, case definitions should ideally include incidents with minor consequences, and near misses (Drupsteen & Guldenmund, 2014; Jacobsson, Ek, & Akselsson, 2012; Lindberg et al., 2010). Near misses have been formally defined as:

- *'A serious error that has potential to cause harm but does not due to chance or interception'* (World Health Organization, p. 10; Leape & Abookire, 2005).

- *'A potential significant event that could have occurred as the consequence of a sequence of actual occurrences but did not occur owing to the plant conditions prevailing at the time'* (p. 127; International Atomic Energy Agency, 2007).

- *'An incident that could have caused serious injury or illness but did not'* (p. 2; Occupational Safety and Health Administration, 2015).

- *'An unintentional unsafe occurrence that could have resulted in an injury, fatality, or property damage. Only a fortunate break in the chain of events prevented an injury, fatality or damage'* (National Fire Fighter Near-Miss Reporting System, p. 556, Taylor & Lacovara, 2015).

The uniting factor amongst these definitions is that near misses are incidents that have the potential to cause harm but did not do so due to various factors. As such, near misses can be a powerful source of information for proactively identifying new hazards, and contributory factors, before they cause harm (Pham, Girard, & Pronovost, 2013). They also provide useful information for the development of incident prevention strategies, as near misses are often prevented via an intervention of some sort.

3.6.3 Data Collection Fields

The data collection fields should be designed to capture information about incidents through an *incident report form*. The system should also be designed to collect comparable exposure data to support the calculation of *incident rates*.

3.6.3.1 Incident Report Form

Incident report forms should not place an undue burden on reporters, as they need to be completed consistently, accurately, and completely. The fields on the form should be minimized, and focus on collecting good quality data on *incident characteristics*, along with a comprehensive description of the *network of contributory factors* involved in the incident.

Information about *incident characteristics* provides a way of filtering data to examine the contributory factors associated with specific types of incidents (e.g. injuries, incidents associated with particular work activities, or types of workers). In general, there are three broad categories of incident characteristics that are useful for filtering data:

1. *Contextual factors* at the time of the incident (e.g. when, where, what type of task or work activity was being undertaken, and who was involved in the task or work activity);
2. *Demographics* of the people impacted by the event (e.g. who was injured, age, gender); and
3. *Outcomes* from the incident (e.g. the nature of injury, the severity of the injury, immediate treatment, evacuation).

The specific questions to include on the incident report form should be determined through consultation with end users to ensure that (a) they make sense to those who will report incidents, (b) the data can be analyzed appropriately and efficiently, and (c) the resulting filters that can be applied to the data are useful for identifying trends and potential problems.

To support the efficient analysis of the data, incident characteristics should generally be collected using standardized response options (i.e. classification fields), rather than free text. To ensure that reporters are not required to force fit their responses into categories, standardized response options need to be designed so that they do not overlap (Wallace & Ross, 2006b). For example, a person reporting a shoulder injury should not be forced to choose between categories such as sprain, strain, dislocation, or tendinitis.

It is also important to ensure that the response options are commonly understood (e.g. day of the week, time, body parts injured), and do not

Standardized response options are utilized to collect data **when appropriate**.

include any technical jargon (e.g. mechanism of injury), or medical terminology (e.g. names of specific musculoskeletal and connective tissue diseases). If the potential response options do not meet these requirements, then free text fields, with appropriate prompts, should be utilized to collect the desired data, or alternatively, the question should not be included on the form.

The majority of the incident report form should focus on collecting information on the *network of contributory* factors involved in the incident. Many incident report forms attempt to collect data on contributory factors using a checklist. This is not recommended as analyses of incident reports using checklists with a narrative description of the incident show: (1) there are often many more contributory factors described in the narrative than the checklist selections indicate; (2) the reasons why reporters have selected a particular factor in the checklist is often unclear from the narrative; and (3) it is difficult to interpret checklist summary data to identify incident prevention strategies without supporting information from the incident reports (Cessford, 2009, 2010; Goode, Salmon, Lenné, & Finch, 2015; Hill, 2011). That is, even when a checklist is used, the narrative section of the reports still needs to be analyzed to identify the full range of contributory factors involved in the incidents.

Consequently, it is recommended that incident reports use free text fields with specific prompts to capture detailed qualitative descriptions of contributory factors and relationships. The prompts should begin by asking for a brief description of the incident to allow the reporter to start thinking about the incident. A series of prompts should then be provided to elicit information regarding the specific contributory factors at each level of the Accimap framework. The prompts should include examples of the relevant contributory factors from the classification scheme. This will help reporters understand how to 'look up and out', rather than 'down and in'. A final prompt should be used to elicit further information about potential relationships between the contributory factors that have been identified. This will help reporters piece together how the contributory factors that they have identified interacted with one another to create the incident in question.

3.6.3.2 Incident Rates

Incident rates indicate the level of exposure to risk for a specific population (e.g. healthcare workers) in a specific context (e.g. working within a specific hospital). Incident rates provide information about how frequently incidents do occur, relative to how frequently they could potentially occur, given the context. For example, the lost time injury frequency rate measures the number of lost-time injuries per million hours worked in a specified time period in a workplace (Worksafe Australia, 1990). The formula is:

$$\frac{\text{Number of lost time injury/disease cases reported in the recording period}}{\text{Number of hours worked in the recording period}} \times 1,000,000$$

As a starting point, two conditions must be met to ensure that incident rates are *representative*. First, there needs to be a clear definition of an incident – this forms the numerator (i.e. the scope of the system as described in Section 3.6.2). Second,

REPRESENTATIVENESS

Incident rates accurately represent how frequently incidents (as defined in the scope) are occurring over time relative to the frequency of exposure.

there needs to be a clear definition of the population at risk – this forms the denominator. This means that in addition to collecting incident reports, comparable information needs to be collected about the population at risk in the specific context (e.g. total number of hours worked, patient days, etc.). It is also useful to collect more specific data on the population at risk relative to specific incident characteristics (e.g. total number of hours worked on specific activities or worksites). This enables more specific incident rates to be calculated, which can be used to evaluate the efficacy of prevention strategies

Most health and safety regulators, and many regulators for specific industries (e.g. rail and aviation), collect data to support industry benchmarking of incident rates. This means that the rates observed in a specific organization can be compared to the rates observed in other organizations in that industry. For example, Safe Work Australia collects lost-time injury and exposure data on industry groups in Australia (https://www.safeworkaustralia.gov.au/topics/lost-time-injury-frequency-rates-ltifr). Different regulators collect different types of exposure data, and recommend different methods for calculating incident rates. The data collection fields for exposure data therefore need to be based on the practices of the relevant regulator in the domain.

It is also important to note that incident rates based on incident reports are not a valid measure of the safety performance of a specific organization, although this is a commonly accepted practice (Safe Work Australia, 2017). Across many domains, studies show that the number of reported incidents severely underestimates the actual number of incidents. Commonly found reasons for under-reporting are fear of retribution, and poor safety climate or culture (Benn et al., 2009; Pham et al., 2013; Probst & Estrada, 2010; Probst & Graso, 2013). So an organization with a relatively low incident reporting rate may well be 'less safe' than one with a much higher reporting rate. Incident rates need to be carefully interpreted in light of these limitations.

3.7 Stage 4: Learning from Incidents Process

The next stage in the development is designing the processes that will be used to collect, analyze, and translate the lessons learned from incident reports into actions.

This involves mapping out the required activities, feedback loops and end users involved in a flowchart. An ideal process for learning from incidents is shown in Figure 3.3. This is based on the literature on learning from incidents (Benn et al., 2009; Drupsteen, Groeneweg, & Zwetsloot, 2013; Jacobsson et al., 2012; Lindberg et al., 2010; Lukic, Littlejohn, & Margaryan, 2012).

The first important feature of this process is that the activities are distributed across everyone within the system (e.g. workers, mangers, executives, etc.). In particular, the people directly involved in the incident, and their supervisors or managers, are required to provide information on incident reports. This will assist in the identification of a more comprehensive set of contributory factors. Similarly, involving everyone in the design of incident prevention strategies will help to generate more holistic strategies, and identify potential unintended consequences of changes (Goode, Read, van Mulken, Clacy, & Salmon, 2016; Rollenhagen, 2011). Regularly involving senior leaders in the process also ensures that there is a sustained commitment to learning from incidents (Barach & Small, 2000).

A second important feature is that feedback loops are used to ensure that *timely, consistent,* and *transparent feedback* is provided to people directly involved in incidents, and key stakeholders. Timely feedback reinforces that safety is a high priority within the organization, and encourages future reporting (Benn et al., 2009).

DATA COLLECTION PROCESS

The method of data collection and the number of steps involved are described in a flowchart.

RESPONSIVE

End users are committed to changing the work system based on the findings.

SUSTAINED LEADERSHIP SUPPORT

Senior leaders are committed to learning from incidents on an ongoing basis.

TIMELINESS

There is rapid turnaround between data collection and reporting meaningful information to stakeholders.

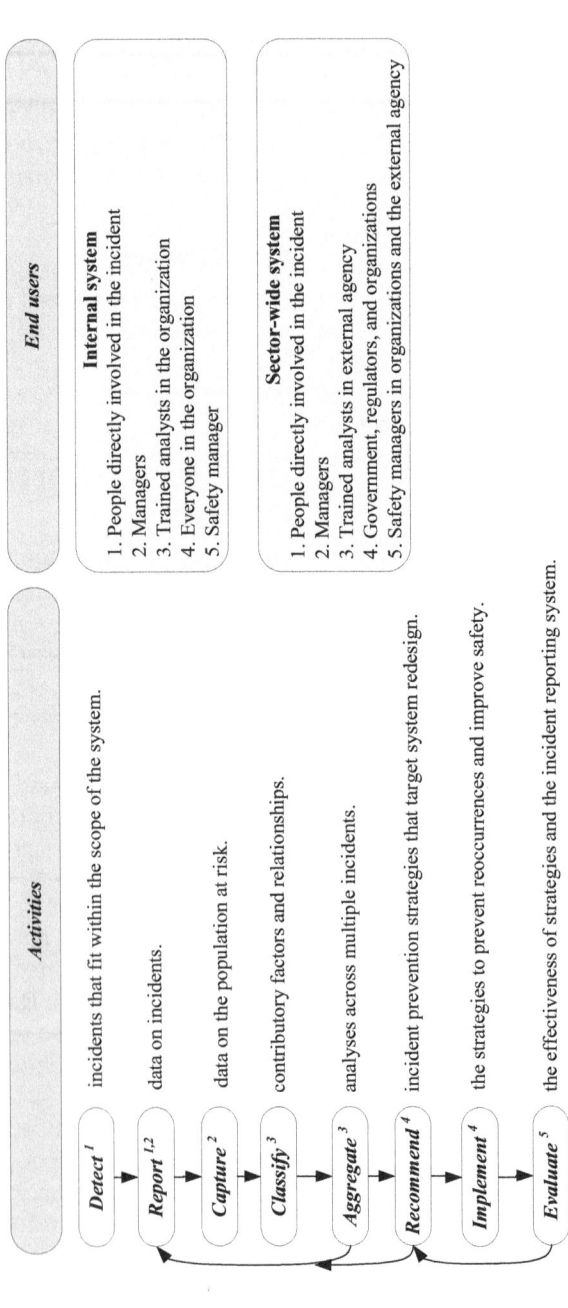

FIGURE 3.3

An ideal process for learning from incidents in sector-wide and internal incident reporting systems, describing the core activities, feedback loops between activities, and the end users who should contribute to each activity.

A third important feature of Figure 3.3 is that *experts* validate the data, and apply the classification scheme. This ensures that the data is *complete*, and that reports are analyzed *accurately* and *consistently*.

A fourth aspect of Figure 3.3 is that, all incident data is aggregated to ensure *data confidentiality and individual privacy*. It is critical that internal systems include safeguards on the reporting of data to protect individual privacy, and avoid opportunities for blame. Similarly, in sector-wide systems, the process for de-identifying the data and constraints on reporting should be clearly defined to protect individuals and organizations.

Finally, the recommendations for incident prevention strategies focus on improving the design of the work system rather than trying to 'fix' specific contributory factors in isolation. The systems thinking approach suggests that to effectively treat specific problems (e.g. errors, staff compliance, or faulty equipment) the system must be designed to prevent their re-occurrence. This involves identifying the interactions across the levels of the system that create these outcomes within the system.

EXPERT ANALYSIS

Reports are validated and evaluated by experts who understand the circumstances, and are trained to analyze the contributory factors involved in the incident.

DATA CONFIDENTIALITY AND INDIVIDUAL PRIVACY

Information that could be used to identify individuals or organizations is not revealed in any outputs or data summaries.

SYSTEMS-ORIENTATED

Incident prevention strategies focus on redesigning the work system to support safe operation.

One additional factor to consider in designing a learning from incidents process for sector-wide systems is whether the data collection process should be *centralized* or *decentralized* (Johnson, 2003). In a centralized system, a reporter (e.g. an employee directly involved in an incident) submits an incident report directly to an external agency. In a decentralized system, the reporter submits the report to an internal incident reporting system within their own organization. The organization then contributes de-identified incident data to an external agency. As there is little evidence to prefer one type of system over the other (Johnson, 2003), the design should be selected based on the needs of the sector.

3.8 Stage 5: Software Tools and Training Materials

The next stage involves developing software tools and training materials to support the implementation of the accident analysis method, data collection protocol, and process for learning from incidents in practice.

The software tools and training materials need to be developed in parallel because they need to be tested and refined as a single product. If end users do not have a sufficient understanding of the underpinning theory and accident analysis method, they will not be able to use the software tool to appropriately report incidents, or to analyze them. The training materials should be embedded within the software tool, so it is easy for end users to understand what they are required to do.

3.8.1 Software Tools

Ideally, software tools should enable:

1. The **collection** of incident and exposure data, using the data collection protocol developed during Stage 3;
2. The **analysis** of the contributory factors, and relationships, identified from incident reports using the contributory factor classification scheme developed during Stage 2;
3. The **storage** of analyzed incident data in an accessible format that makes it easy to identify relevant cases and produce aggregate summary reports;
4. The **reporting** of statistics on incident characteristics, rates, and aggregate analyses of contributory factors; and
5. The **documentation**, and regular updating, of incident prevention strategies, implementation, and evaluation.

As technology changes rapidly, it would be unwise to specify the use of any particular software or database structure for an incident reporting system (Johnson, 2002). There are, however, several characteristics that have been identified as optimal features of incident reporting software tools, as shown in Table 3.1. The *practical* characteristics are factors influencing the successful operation of an incident reporting system in an organization (i.e. by improving reporting rates, or minimizing workload). The *operational* characteristics are factors influencing the quality, reliability, and trustworthiness of the data, and resulting analyses. These characteristics can be used as criteria to select an appropriate platform for an incident reporting software tool.

3.8.2 Training Materials

The training materials should describe:

- The underpinning theory (e.g. the principles of the systems thinking approach, Rasmussen's risk management framework and Accimap);
- The agreed process for learning from incidents in the organization;

TABLE 3.1

Optimal Characteristics of Incident Reporting Software Tools

Characteristic	Definition
Practical Characteristics	
Accessibility	Data are stored in a format that makes it easy to identify relevant cases and produce aggregate summary reports.
Availability	The system is readily available to anyone wishing to report an incident.
Ease of reporting	Submitting an incident report is easy (e.g. the report can be completed in the field; reporters are able to save and complete report later).
Sustainability	The system is easy to maintain and update.
Simplicity	The database for storing information has a simple structure and is easy to operate.
Utility	The tool is practical, affordable, and does not put unnecessary workload on reporters or organizations.
Operational Characteristics	
Quality control measures	Data validation is used to ensure that data are entered into the tool accurately.
Credible	The data analyses and summaries can be relied upon as accurate reflections of the data.
Flexibility	The system is easy to change, especially when evaluation shows that change is necessary or desirable.
Stability of the system	The system is reliably available for data input at all times.
System security	Data access is controlled to safeguard against disclosure of confidential material.

- How to report an incident, including the scope of the system and potential contributory factors and relationships that may play a role;

- How to analyze incident reports using the contributory factor classification scheme and produce aggregate analyses of multiple incidents; and

- How to interpret the aggregate analyses to identify appropriate incident prevention strategies that target system redesign.

COMPREHENSIVE GUIDANCE

Training materials support the appropriate reporting, analysis and, interpretation of incident data.

The training material should be embedded into the software tool so that end users can access the relevant information as required. Additional training materials may also need to be developed based on end users' familiarity with the systems thinking approach. Ongoing training and coaching is usually required to ensure that everyone in the organization has a sufficient understanding of systems thinking.

3.8.3 End-User Testing

A core requirement for the software tool and training materials is satisfactory *usability*. Usability is the extent to which specified end users are able to achieve their goals *effectively, efficiently*, and with a feeling of *satisfaction* (ISO 9241-11, 1998).

> **USABILITY**
>
> End users can achieve their goals effectively, efficiently, and with a feeling of satisfaction.

Usability can only be evaluated by conducting assessments with end users. Early in the development process, multiple rounds of informal feedback should be used to determine whether proposed designs fit the context, and identify any problems. Once the prototypes have been refined (i.e. they are at a stage where they are ready to be deployed in an organization), summative usability assessments should be used to evaluate the incident reporting system to identify areas for improvement. Chapter 9 describes how to undertake this type of assessment.

3.9 Stage 6: Implementating-Evaluating Data Quality

The final stage of the process model is implementing the incident reporting system, and evaluating the *completeness and validity of recorded data* (German et al., 2001). It is good practice to initially conduct a small-scale implementation trial to identify any issues with data quality prior to full-scale implementation, and then periodically evaluate data quality throughout the life cycle of the incident reporting system.

There are five characteristics that are relevant to assessing the quality of data collected via an incident reporting system, shown in Table 3.2. Chapter 10 describes how to design and conduct an implementation trial to evaluate these characteristics.

TABLE 3.2

Data Quality Characteristics for Incident Reporting Systems Underpinned by Systems Thinking

Characteristic	Definition
Data completeness	A consistent amount of data is provided about every reported incident.
Positive predictive value	The incident report provides an accurate description of the incident.
Sensitivity	All relevant incidents that occur are reported.
Specificity	No irrelevant cases are reported.
Representativeness	The incident rates accurately represent how frequently incidents (as defined in the scope) are occurring over time relative to the frequency of exposure.

3.10 Criteria for Designing and Evaluating Incident Reporting Systems

This chapter has presented a process model for developing an incident reporting system, underpinned by the principles of systems thinking, and good practice for incident and injury data collection, and learning from incidents. The characteristics that have been identified throughout this chapter are the hallmarks of a well-designed and effectively implemented incident reporting system. They can therefore be used as criteria to guide the design of new incident reporting systems, and and to evaluate existing ones. The stages of the process model and relevant criteria are summarized below.

3.10.1 Stage 1: Understand the Context

1. **Acceptability:** The system matches the characteristics, needs and priorities of end users.

3.10.2 Stage 2: Accident Analysis Method

1. **Usefulness:** The scheme includes sufficiently detailed classifications so that aggregate analyses of incidents can easily be interpreted.
2. **Reliability:** The same end user classifies contributory factors in the same way on repeated occasions, and different end users classify contributory factors in the same way.
3. **Validity:** End users classify contributory factors in line with expert opinion.

3.10.3 Stage 3: Data Collection Protocol

1. **Clear purpose and objectives:** There is a clearly defined reason why the incident reporting system exists and how the data are to be used.
2. **Clear case definitions:** There are clear definitions of what should and should not be reported.
3. **Appropriate use of standardized response options:** Standardized response options are utilized to collect data when appropriate.
4. **Data confidentiality and individual privacy:** Information that could identify individuals or organizations are not revealed in any outputs or data summaries.
5. **Non-punitive:** Reporters are free from the fear of retaliation or punishment as the result of reporting.
6. **Representativeness:** Incident rates accurately represent how frequently incidents (as defined in the scope) are occurring over time, relative to the frequency of exposure.

3.10.4 Stage 4: Learning from Incidents Process

1. **Data collection process:** The method of data collection, and the number of steps involved, are described in a flowchart.
2. **Expert analysis:** Reports are validated and evaluated by experts who understand the circumstances, and are trained to analyze the contributory factors involved in the incident.
3. **Responsive:** End users are committed to changing the work system based on the findings.
4. **Timeliness:** There is rapid turnaround between data collection and reporting meaningful information to stakeholders.
5. **Sustained leadership support:** Senior leaders are committed to learning from incidents on an ongoing basis.
6. **Systems-orientated:** Incident prevention strategies focus on redesigning the work system to support safe operation.

3.10.5 Stage 5: Software Tools and Training Materials

1. **Accessibility:** Data are stored in a format that makes it easy to identify relevant cases and produce aggregate summary reports.
2. **Availability:** The system is readily available to anyone wishing to report an incident.
3. **Comprehensive guidance:** Training materials support the appropriate reporting, analysis, and interpretation of incident data.
4. **Credibility:** The data analyses and summaries can be relied upon as accurate reflections of the data.
5. **Ease of reporting:** Submitting an incident report is easy.
6. **Flexibility:** The system is easy to change, especially when evaluation shows that change is necessary or desirable.
7. **Quality control measures:** Data validation is used to ensure that data are entered into the tool accurately.
8. **Simplicity:** The database for storing information has a simple structure and is easy to operate.
9. **Stability of the system:** The system is reliably available for data input at all times.
10. **Sustainability:** The system is easy to maintain and update.
11. **System security:** Data access is controlled to safeguard against disclosure of confidential material.
12. **Usability:** End users can achieve their goals effectively, efficiently, and with a feeling of satisfaction.
13. **Utility:** The system is practical, affordable, and does not put unnecessary workload on reporters or organizations.

3.10.6 Stage 6: Implementation

1. **Data completeness:** A consistent amount of data is provided about every reported incident.
2. **Positive predictive value:** The incident report provides an accurate description of the incident.
3. **Representativeness:** The data accurately represents how frequently incidents, and the contributory factors involved, are occurring over time.
4. **Sensitivity:** All relevant incidents that occur are reported.
5. **Specificity:** No irrelevant cases are reported.

3.11 Next Steps

The following chapters provide step-by-step guidance on using the process model presented in this chapter to develop a new incident reporting system. The process is illustrated by presenting the development of the Understanding and Preventing Led Outdoor Accidents Data System (UPLOADS) as a case study. UPLOADS was designed as a decentralized sector-wide incident reporting system, where individual organizations collect and analyze their own data and contribute de-identified data to a National Incident Dataset. The case study is therefore relevant to readers wishing to develop an internal incident reporting system, or a sector-wide incident reporting system.

References

Baber, C. (2002). Subjective evaluation of usability. *Ergonomics, 45*(14), 1021–1025. doi:10.1080/00140130210166807

Barach, P., & Small, S. D. (2000). Reporting and preventing medical mishaps: Lessons from non-medical near miss reporting systems. *BMJ: British Medical Journal, 320*(7237), 759.

Benn, J., Koutantji, M., Wallace, L., Spurgeon, P., Rejman, M., Healey, A., & Vincent, C. (2009). Feedback from incident reporting: Information and action to improve patient safety. *Quality and Safety in Health Care, 18*(1), 11–21.

Cessford, G. (2009). *National Incident Database report 2007–2008: Outdoor education and recreation.* Retrieved from https://www.incidentreport.org.nz/resources/NID_Report_2007-2008.pdf Access date: 15th May 2018.

Cessford, G. (2010). *National Incident Database report 2009: Outdoor education and recreation* Retrieved from https://www.incidentreport.org.nz/resources/NID_Report_2010.pdf. Access date: 15th May 2018.

Dekker, S. (2011). *Drift into failure: From hunting broken components to understanding complex systems*. Boca Raton, FL: CRC Press, Taylor & Francis Group.

Dekker, S. (2012). *Just Culture: Balancing Safety and Accountability*. Retrieved from http://MONASH.eblib.com.au/patron/FullRecord.aspx?p=906963

Drupsteen, L., Groeneweg, J., & Zwetsloot, G. (2013). Critical steps in learning from incidents: Using learning potential in the process from reporting an incident to accident prevention. *International Journal of Occupational Safety and Ergonomics, 19*(1), 63–77. doi:10.1080/10803548.2013.11076966

Drupsteen, L., & Guldenmund, F. W. (2014). What is learning? A review of the safety literature to define learning from incidents, accidents and disasters. *Journal of Contingencies and Crisis Management, 22*(2), 81–96. doi:10.1111/1468-5973.12039

German, R., Lee, L., Horan, J., Milstein, R., Pertowski, C., & Waller, M. (2001). Guidelines Working Group, Centers for Disease Control and Prevention (CDC). Updated guidelines for evaluating public health surveillance systems: Recommendations from the Guidelines Working Group. *MMWR Recomm Rep, 50*(RR-13), 1–35.

Goode, N., Finch, C., Cassell, E., Lenné, M. G., & Salmon, P. M. (2014). What would you like? Identifying the required characteristics of an industry-wide incident reporting and learning system for the led outdoor activity sector. *Australian Journal of Outdoor Education, 17*(2), 2–15.

Goode, N., Read, G. J., van Mulken, M. R., Clacy, A., & Salmon, P. M. (2016). Designing system reforms: Using a systems approach to translate incident analyses into prevention strategies. *Frontiers in Psychology, 7*.

Goode, N., Salmon, P. M., Lenné, M., & Finch, C. F. (2015). *Looking beyond people, equipment and environment: Is a systems theory model of accident causation required to understand injuries and near misses during outdoor activities?* Paper presented at the 6th International Conference on Applied Human Factors and Ergonomics (AHFE 2015), Las Vegas, U.S.

Hewitt, T. A. (2013). Incident reporting systems – The hidden story. *Safety and Reliability, 33*(2), 13–28. doi:10.1080/09617353.2013.11716252

Hill, A. (2011). *National Incident Database report 2010: Outdoor education and recreation*. Retrieved from https://www.incidentreport.org.nz/resources/NID _Report_2010.pdf. Access date: 19th June 2018.

International Atomic Energy Agency. (2007). *IAEA safety glossary: Terminology used in nuclear safety and radiation protection*: Internat. Atomic Energy Agency.

Jacobsson, A., Ek, A., & Akselsson, R. (2011). Method for evaluating learning from incidents using the idea of 'level of learning'. *Journal of Loss Prevention in the Process Industries, 24*(4), 333–343. doi:http://dx.doi.org/10.1016/j.jlp.2011 .01.011

Jacobsson, A., Ek, A., & Akselsson, R. (2012). Learning from incidents – A method for assessing the effectiveness of the learning cycle. *Journal of Loss Prevention in the Process Industries, 25*(3), 561–570. doi:http://dx.doi.org/10.1016/j.jlp.2011.12.013

Johnson, C. (2002). Software tools to support incident reporting in safety-critical systems. *Safety Science, 40*(9), 765–780. doi:http://dx.doi.org/10.1016/S0925-7535 (01)00085-6

Johnson, C. W. (2003). *Failure in safety-critical systems: A handbook of incident and accident reporting*. Scotland: University of Glasgow.

Leape, L. L., & Abookire, S. (2005). *WHO draft guidelines for adverse event reporting and learning systems: From information to action*. World Health Organization.

Lindberg, A. K., Hansson, S. O., & Rollenhagen, C. (2010). Learning from accidents – What more do we need to know? *Safety Science, 48*(6), 714–721.

Lukic, D., Littlejohn, A., & Margaryan, A. (2012). A framework for learning from incidents in the workplace. *Safety Science, 50*(4), 950–957. doi:http://dx.doi.org /10.1016/j.ssci.2011.12.032

Lundberg, J., Rollenhagen, C., & Hollnagel, E. (2009). What-you-look-for-is-what-you-find – The consequences of underlying accident models in eight accident investigation manuals. *Safety Science, 47*(10), 1297–1311. doi:http://dx.doi.org/10.1016/j .ssci.2009.01.004

Maguire, M. (2001). Context of use within usability activities. *International Journal of Human-Computer Studies, 55*(4), 453–483.

Nielsen, J. (1994). *Usability engineering*: Elsevier.

Occupational Safety and Health Administration. (2015). *Incident [accident] investigations: A guide for employers*. Retrieved from https://www.osha.gov/dte/Inc InvGuide4Empl_Dec2015.pdf.

Pham, J. C., Girard, T., & Pronovost, P. J. (2013). What to do with healthcare incident reporting systems. *Journal of Public Health Research, 2*(3).

Probst, T. M., & Estrada, A. X. (2010). Accident under-reporting among employees: Testing the moderating influence of psychological safety climate and supervisor enforcement of safety practices. *Accident Analysis and Prevention, 42*(5), 1438–1444. doi:https://doi.org/10.1016/j.aap.2009.06.027

Probst, T. M., & Graso, M. (2013). Pressure to produce = pressure to reduce accident reporting? *Accident Analysis and Prevention, 59C*, 580–587.

Reason, J. (1997). *Managing the risks of organizational accidents*. Aldershot, UK: Ashgate.

Rollenhagen, C. (2011). Event investigations at nuclear power plants in Sweden: Reflections about a method and some associated practices. *Safety Science, 49*(1), 21–26. doi:http://dx.doi.org/10.1016/j.ssci.2009.12.012

Safe Work Australia. (2017). Lost time injury frequency rates (LTIFR). Retrieved from https://www.safeworkaustralia.gov.au/statistics-and-research/lost-time -injury-frequency-rates-ltifr

Salmon, P. M., Cornelissen, M., & Trotter, M. J. (2012). Systems-based accident analysis methods: A comparison of Accimap, HFACS, and STAMP. *Safety Science, 50*(4), 1158–1170.

Salmon, P. M., Stanton, N. A., Lenné, M., Jenkins, D. P., Rafferty, L., & Walker, G. H. (2011). *Human factors methods and accident analysis practical guidance and case study applications*. UK: Ashgate.

Sanne, J. M. (2008). Incident reporting or storytelling? Competing schemes in a safety-critical and hazardous work setting. *Safety Science, 46*(8), 1205–1222. doi:http:// dx.doi.org/10.1016/j.ssci.2007.06.024

Sherman, H., Loeb, J., & Fletcher, M. M. (2005). Project to develop the international patient safety event taxonomy: Updated review of the literature 2003–2005. *Proceedings of The WHO World Health Organization Alliance for Patient Safety*.

Svedung, I., & Rasmussen, J. (2002). Graphic representation of accident scenarios: Mapping system structure and the causation of accidents. *Safety Science, 40*(5), 397–417.

Taylor, J. A., & Lacovara, A. V. (2015). From infancy to adolescence: The development and future of the national firefighter near-miss reporting system. *NEW SOLUTIONS: A Journal of Environmental and Occupational Health Policy, 24*(4), 555–576. doi:10.2190/NS.24.4.h

Underwood, P., & Waterson, P. (2013). Accident Analysis Models and Methods: Guidance for Safety Professionals.

Usability.gov. (2018). User-centered design basics. Retrieved from www.usability.gov

Vaughan, D. (1997). *The Challenger launch decision: Risky technology, culture, and deviance at NASA*: University of Chicago Press.

Wallace, B., & Ross, A. J. (2006a). *Beyond human error: Taxonomies and safety science*. USA: CRC Press.

Wallace, B., & Ross, A. J. (2006b). Safety and taxonomies. *Beyond human error: Taxonomies and safety science* (pp. 31–64): CRC Press.

Worksafe Australia. (1990). Workplace injury and disease recording standard. *Worksafe Australia*.

4

Understanding the Context

Practitioner Summary

This chapter establishes the context for developing the Understanding and Preventing Led Outdoor Accidents Data System (UPLOADS), and illustrates the processes required to evaluate whether existing knowledge on incident data collection and causation in a domain aligns with systems thinking. The chapter begins by describing the problems with the approach to incident data collection in Australia prior to the development of UPLOADS. It then describes the review that was conducted to evaluate whether existing knowledge in the domain aligned with systems thinking. The core conclusions from this review were: (1) existing incident reporting systems limited data collection and analysis to the immediate context and (2) existing domain-specific accident causation models focused almost exclusively on people, equipment, and the environment as the primary causes of incidents. This clearly established the need for developing a new incident reporting system, underpinned by systems thinking, for the domain.

Practical Challenges

Many domains have not developed domain-specific accident causation models or sector-wide incident reporting systems. In these domains, it would be more appropriate to evaluate whether the accident causation models and incident reporting systems that are used in practice by relevant organizations align with systems thinking.

4.1 Introduction

Prior to developing a new incident reporting system underpinned by systems thinking, it is important to establish (a) whether there is already an appropriate incident reporting system; or (b) whether there is existing knowledge on incident data collection and causation in the domain that could inform the development of the new system. This will ensure that the development does not unnecessarily replicate existing work.

The aim of this chapter is to illustrate a process for evaluating whether existing incident reporting systems, contributory factor classification schemes, and accident causation models align with systems thinking. The chapter also serves to introduce the reader to the context of use for the development of the incident reporting system that is described in the rest of this book, UPLOADS.

4.2 The Context for This Case Study

'Led outdoor activities' (LOA) are formally defined as facilitated or instructed activities within outdoor education and recreation settings (Carden, Goode, & Salmon, 2017; Salmon, Williamson, Lenné, Mitsopoulos-Rubens, & Rudin-Brown, 2010). Examples include bushwalking, canyoning, kayaking, rock climbing, and camping, to name only a few. A key feature of these activities is that they involve intentional engagement with physical risk. Organizations that provide LOAs (LOA providers) also have a duty of care towards those involved in their activities (e.g. leaders, participants, volunteers) to eliminate or manage the risks involved as far as reasonably practicable (Goode, Finch, Cassell, Lenné, & Salmon, 2014).

In Australia, LOA providers include a diverse range of organizations, including not-for-profit outdoor education and recreation providers, outdoor education departments within schools, school camps, adventure tourism operators and outdoor therapy programs. These organizations range in size from large, with many hundreds of staff, to sole operators. They are also distributed across Australia, often in remote areas (Service Skills Australia, 2010, 2013; Williams & Allen, 2012).

In contrast to outdoor activities undertaken for personal recreation, the provision of LOAs involves several clearly defined actors and organizations. Figure 4.1 presents those that play a role in managing safety during LOAs in Australia, classified according to the levels of Rasmussen's (1997) risk management framework.

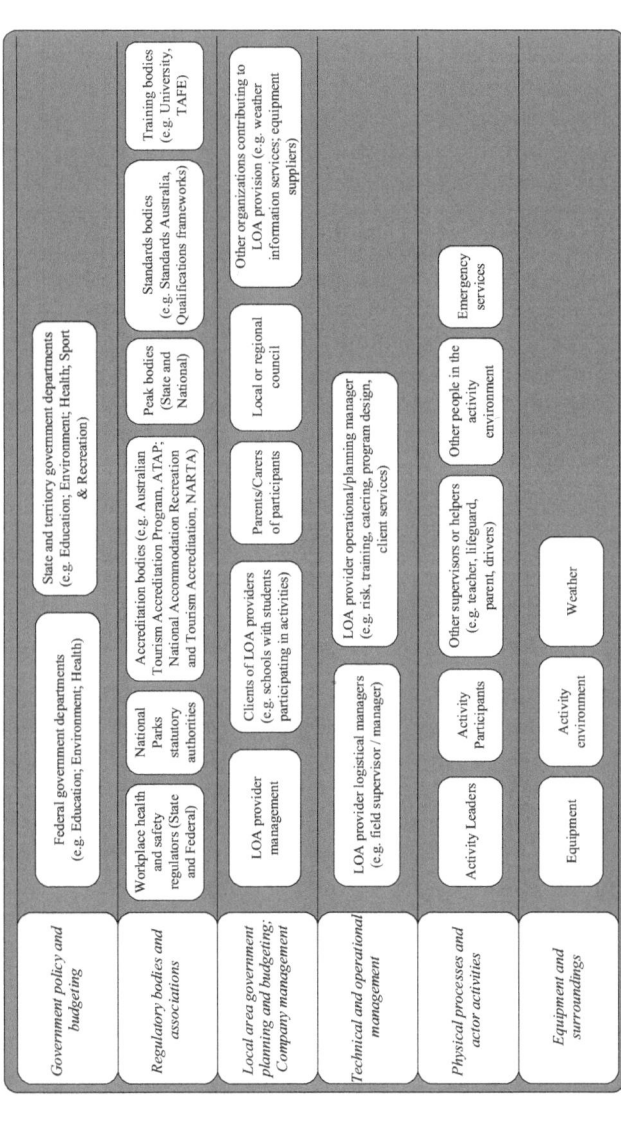

FIGURE 4.1
The actors and organizations that play a role in managing safety during LOAs in Australia, presented according to the levels of Rasmussen's 1997 risk management framework.

The bottom two levels of Figure 4.1 represents the actors at the 'sharp end' of the system: activity participants perform LOAs under the instruction and supervision of an activity leader, who is often supported by other supervisors and helpers. There may also be other people, external to the group, in the activity environment who impact the conduct of the LOA program (e.g. members of the public).

Figure 4.1 also illustrates that there are many actors and organizations who contribute to managing the safety of LOA programs outside of the immediate context of the activity environment. Within LOA providers, there are typically several people involved in the scheduling, design, and management of LOA programs. Schools with students participating in activities are usually responsible for collecting information such as medical forms and dietary requirements to inform the design of LOA programs. Parents or carers with children participating in activities are responsible for providing accurate information about their child, and ensuring that the child has adequate equipment and clothing for the trip. All LOA providers must comply with the relevant workplace safety legislation and regulation. There are also many peak bodies, authorities, and government departments that require LOA providers to meet certain safety standards during the delivery of LOAs (Goode et al., 2014).

4.3 Injury and Incident Data Collection in Australia Prior to UPLOADS

Just like any other organization, LOA providers require detailed information on incident characteristics and causation to appropriately manage the risk associated with their work activities. In Australia, however, prior to embarking on the UPLOADS research program, there was a paucity of published research on LOA incidents (Dickson, 2012). The few published studies focused on fatalities reported in newspapers and coroner's inquests (Brookes, 2003, 2007a,b, 2011). This potentially presented a biased view of the nature of LOA incidents in Australia, as studies conducted in other countries indicated that most injuries are relatively minor (Barst, Bialeschki, & Comstock, 2008; Hamonko, McIntosh, Schimelpfenig, & Leemon, 2011; Leemon & Schimelpfenig, 2003). As a result, there were repeated calls from within the Australian sector for a national approach to LOA incident reporting to support evidence-based decisions on participation, funding, and risk management (Brackenberg, 1999; Brown & Fraser, 2009; Dickson, 2012).

Prior to the development of UPLOADS, many LOA providers had developed their own internal incident reporting systems. For example, Outward Bound Australia developed the Outdoor Medical Incident Database (Salmon et al., 2009). Independent schools, such as St Joseph's College, which operates an Outdoor Education Centre north of Sydney, had also developed a reporting system for capturing specific information on injuries and near misses (Brackenberg, 1997). Little information, however, was available regarding the actual prevalence of internal incident reporting systems within the sector, their quality, or their impact on safety.

The only overarching scheme that supported the collection of data from all LOA providers (and still does), was the workplace health and safety (WHS) regulatory system in each Australian state. These systems require LOA providers to report notifiable injuries (i.e. those requiring treatment as an in-patient of a hospital) and fatalities. A significant limitation of these systems are that they do not provide a mechanism for reporting and analyzing less severe incidents. Consequently, the data presents a biased view of incidents during LOAs (Goode et al., 2014).

There were (and still are) several injury surveillance systems that potentially captured data on hospital admissions and fatalities during LOA programs. For example, the National Coronial Information System, the Victorian State Trauma Registry, the Victorian Injury Surveillance System, the Admitted Patient Data Collection, the Queensland Injury Surveillance Unit, and Kidsafe Western Australia. Although LOA injuries are undoubtedly captured by these systems, retrieving data on injuries associated with LOA activities is difficult. Many of these systems are based on the International Classification of Diseases (ICD), in which over 200 sports and leisure activity codes can be used, but many relevant cases are typically not classified (Finch & Boufous, 2008). Moreover, the ICD activity codes do not include codes for many of the activities which form the core of LOA (e.g. camping and high ropes). Finally, the ICD-activity codes do not distinguish between activities that take place within the context of a facilitated and managed program, and those that are undertaken independently (Mitchell, Boufous, & Finch, 2008). Extracting relevant cases on incidents during LOAs from these injury surveillance systems is consequently highly resource intensive and impractical on a regular basis.

The fragmented approach to incident data collection across LOA providers, and lack of comprehensive injury data, indicated that there was a need for a nationwide, standardized approach to incident reporting in the LOA sector. Before, before UPLOADS there had been several attempts to develop anonymous or confidential incident reporting systems within the Australian LOA sector. The Australian Accident Register (AAR) was established in 2007 to collect anonymous reports from individuals involved in incidents occurring in outdoor and adventure environments (Salmon et al., 2009); it was closed in early 2011. The National Database of Accidents and Incidents in Outdoor Programs collected voluntary reports on incidents occurring during outdoor education programs in Australia from 1995 to 1999 (Dickson, Chapman, & Hurrell, 2000). The U.S.-based Wilderness Risk Management Committee (WRMC) Adventure Program Incident Report Project also maintained an Australian database for a short period of time (Brackenberg, 1997, 1999). While these incident reporting systems ultimately closed, their short periods of success indicated that the potential benefits of sharing incident data was well known in Australia's LOA sector prior to the development of UPLOADS.

4.4 State of Knowledge in the LOA Domain

The following sections summarize the findings from a literature review that was undertaken prior to the development of UPLOADS in 2011 (Sections 4.1 and 4.2). The purpose of the review was to evaluate whether there were existing LOA-specific incident reporting systems, contributory factor classification schemes, and accident causation models that aligned with systems thinking that could inform (or replace) the development of UPLOADS. The findings are presented to illustrate a process for evaluating existing knowledge in a domain for readers wishing to develop their own incident reporting system.

4.4.1 LOA Incident Reporting Systems and Classification Schemes

The review found that several sector-wide incident reporting systems had been developed for the LOA domain. The review evaluated the types of information they collect, and their contributory factor classification schemes.

4.4.1.1 *National Outdoor Leadership School (NOLS) Database (1978–Present)*

NOLS in North America has recorded wilderness injuries and evacuations in a database since 1978, and regularly reports on the data (Leemon & Schimelpfenig, 2003). The system collects information on the injury itself, the activity involved, and any contributory factors. The classification of contributory factors is limited to the actions and experience of the activity participant, the environment, and the direct mechanism of injury (e.g. fall or blunt trauma) (Hamonko et al., 2011; Leemon & Schimelpfenig, 2003).

4.4.1.2 *The Wilderness Risk Managers Committee (WRMC) Incident Reporting Project (1992–2009)*

Established in 1992, the WRMC Incident Reporting Project involved a consortium of outdoor schools and organizations in the United States working towards a better understanding and management of risks in the outdoors. A standardized form was used to gather information on the type and location of the injury or illness, medical information about the person impacted (e.g. vital signs and history), a narrative description of the incident, with an analysis of contributory factors. The analysis involved ranking the importance of 37 contributory factor categories describing: planning for the activity; the people involved in the activity; the environment; and equipment (Leemon, 2009).

4.4.1.3 New Zealand National Incident Database (NZ NID) (2005 to Present)

The Mountain Safety Council developed the NZ NID to provide a standardized approach to incident reporting for the New Zealand outdoor sector. There are two distinct databases: the Outdoor Education/Recreation NID and the Snow Sports NID. The former database is intended to capture information on outdoor recreation activities conducted by commercial, school-based, not-for-profit, and informal outdoor education and recreation groups (Cessford, 2009). Each report includes: an assessment of incident characteristics; a description of the incident and contributory factors; and a checklist for classifying the contributory factors. The checklist includes four broad categories of contributory factors (activity leader, activity participants, equipment, and environment), with detailed subcategories (Hill, 2011).

4.4.1.4 The Australian Accident Register (2006–2011)

The Australian Accident Register (AAR) was established by a group of volunteers as a repository for sharing the lessons learned from incidents that occurred in the outdoors. All incidents were reported on a voluntary basis. The incident form includes a classification field for activity type and a narrative description of the event. No contributory factor categories were included on the form (Salmon et al., 2009).

4.4.1.5 Healthy Camps Study: CAMP RIO™
(Reporting Information Online) (2006–2010)

The American Camp Association conducted a five-year study of injuries and illnesses in day and resident camps using an online data entry system called CAMP RIO™ (Reporting Information Online). The system collected information about the person involved, when and where the incident happened, the type of injury or illness, and the incident context. The contributory factors categories were limited to the direct mechanism of injury, weather influences, use of protective equipment, and whether the people involved had participated in formal safety training (American Camp Association, 2011).

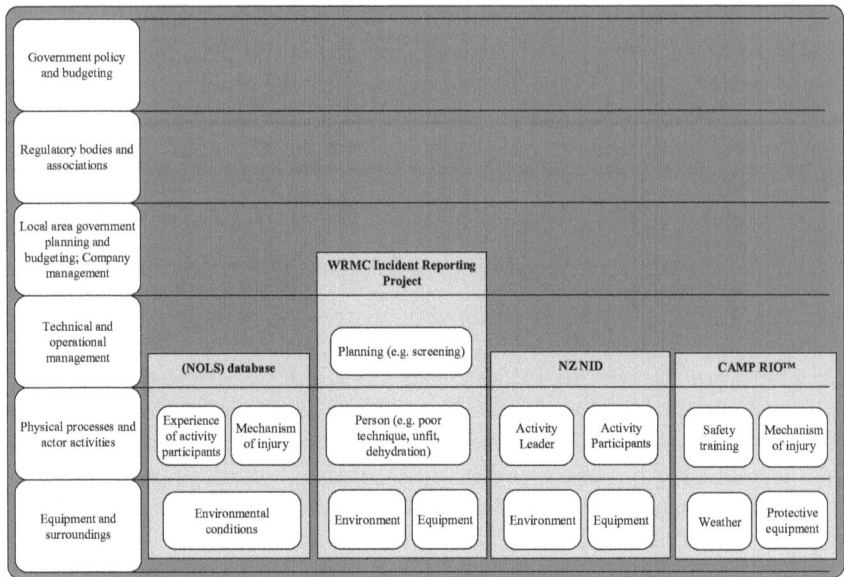

FIGURE 4.2
Summary of the types of contributory factors considered in LOA incident reporting systems mapped onto the levels of Rasmussen's Accimap.

4.4.1.6 *Do These Incident Reporting Systems Support Systems Thinking?*

Although all of these incident reporting systems capture qualitative information about the incident (which could be used to undertake a system analysis), none of them explicitly support a consideration of factors at the higher levels of Rasmussen's risk management framework. This is demonstrated in Figure 4.2, which clearly shows that almost all of the existing systems limit analyses to the immediate context of the activity. Only the WMRC system supported a consideration of factors pre-dating the incident, and this is limited to a single category (planning). Based on this analysis, it was concluded that an appropriate incident reporting system did not already exist, and that the development of a new incident reporting system was required.

4.4.2 LOA Accident Causation Models

The review found that several accident causation models had been developed specifically for the LOA domain. The following points were used to evaluate the alignment of the models with systems thinking:

- Theory: Describes any underpinning models or theoretical frameworks (e.g. systems theory, complexity theory, control theory, information processing theory);

- Contributory factor categories: Describe the categories of contributory factors identified by the model (e.g. leader, environment, participant, organizational);
- Hierarchical system structure: Whether the model represents the structure of the LOA system as a hierarchy of sub-systems;
- Multiple factors and actors: Whether the model proposes that multiple factors and actors contribute to incident causation; and
- Interactions: Whether the model proposes that incidents are caused by interactions between contributory factors or components within the system.

The findings from the evaluation are shown in Table 4.1. Although none of the existing models are explicitly underpinned by systems thinking, they do reflect some of the core principles. All of the models are underpinned by the idea that incidents are caused by multiple contributory factors. Most specify that accidents are caused by interactions between factors. The key gap, however, is a focus on the immediate context of the incident; contributory factors are thought to relate only to the activity, equipment, the leaders and participants, and the activity environment. Only two models consider the role of organizational factors (i.e. Pathways to Change model and the Root Cause model). None of the models consider higher level system factors (e.g. regulators and government) As discussed in Chapter 1, failing to consider factors at these levels is likely to result in the development of human-focused incident prevention strategies that do not treat many of the underlying systemic causes of unsafe behavior (Dekker, 2002; Reason, 1997). Based on this analysis, it was concluded that existing LOA accident causation models were insufficient to underpin the development of a contributory factor classification scheme for a systems thinking–based incident reporting system (Salmon et al., 2009).

> Prior to UPLOADS, existing accident causation models largely focused on the immediate context of the incident and were not well aligned with systems thinking.

4.5 Summary and Next Steps

In summary, prior to the development of UPLOADS, the major barriers to understanding and preventing incidents in the LOA sector were identified as: (1) a lack of a standarized approach to incident reporting and analysis in the Australian LOA sector; and (2) little good quality data on incidents during LOA was publicly available apart from Coroner's reports. The UPLOADS Project was initiated by our industry partners to address these gaps.

TABLE 4.1

Description and Evaluation of Existing LOA Accident Causation Models

Model	Theory	Contributory Factor Categories	Hierarchical System Structure	Multiple Factors	Interactions
Dynamics of accidents model (Hale, 1984 as cited in Curtis 1995)	None	Environmental hazards (environment, equipment, transportation); human factor hazards (participants, leaders, drivers, group)	No	Yes	Yes
Accident matrix model (Meyer, 1979)	None	Unsafe conditions (environment), unsafe acts (clients), judgement errors (instructors)	No	Yes	Yes
Fields of attention in the safety of outdoor activities model (Hovelynck, 1998)	Adapted from organizational and behavioral processes model	Product (goals, techniques), procedure (structuring, assigning roles); and process (personality, emotions)	No	Yes	Yes
Pathways to change model (Haddock, 1999)	Loss causation model (Bird & Germain, 1992)	Lack of management control (inadequate program, standards and compliance); basic causes (personal and job factors); immediate causes (substandard acts and conditions)	No	Yes	Yes
Root cause model (Davidson, 2007)	Reason (1990)	Inadequate safety management systems, poor implementation of safety management systems, unsafe environment, unsafe clients, unsafe equipment, instructor errors	No	Yes	Yes
Behavioral risk management and cascade of events models (Gray & Moeller, 2010)	None	Technical, behavioral, and physical risk factors	No	Yes	No

The review and evaluation conducted during the first phase of the development of UPLOADS clearly established that the existing knowledge on incident data collection and causation was not well aligned with systems thinking. This meant that all the components of UPLOADS (i.e. the classification scheme, data collection protocol, software tool, and training material) had to be designed, almost from the ground up, to align with systems thinking.

For readers wishing to develop an incident reporting system in their own context, it is recommended that a similar review is conducted prior to developing a new system. This is to ensure that work is not unnecessarily replicated.

The next chapter describes the processes required to identify the needs and priorities of end users for the development of a new incident reporting system.

References

American Camp Association. (2011). *Healthy camp study impact report 2006–2010: Promoting health and wellness among youth and staff through a systematic surveillance process in day and resident camps.* Retrieved from http://www.acacamps.org/sites/default/files/images/education/Healthy-Camp-Study-Impact-Report.pdf

Barst, B., Bialeschki, M. D., & Comstock, R. D. (2008). Healthy camps: Initial lessons on illnesses and injuries from a longitudinal study. *Journal of Experiential Education, 30*(3), 267–270.

Brackenberg, M. (1997). How safe are we? A review of injury and illness in outdoor education programmes. *Journal of Adventure Education and Outdoor Leadership, 14*(1).

Brackenberg, M. (1999). Learning from our mistakes – Before it's too late. *Australian Journal of Outdoor Education, 3*(27–33).

Brookes, A. (2003). Outdoor education fatalities in Australia 1960–2002. Part 1. Summary of incidents and introduction to fatality analysis. *Australian Journal of Outdoor Education, 7*(2), 20–35.

Brookes, A. (2007a). Preventing death and serious injury from falling trees and branches. *Australian Journal of Outdoor Education, 11*(2), 50–59.

Brookes, A. (2007b). Research update: Outdoor education fatalities in Australia. *Australian Journal of Outdoor Education, 11*(1), 3–9.

Brookes, A. (2011). Research update 2010: Outdoor education fatalities in Australia. *Australian Journal of Outdoor Education, 15*(1), 35.

Brown, M., & Fraser, D. (2009). Re-evaluating risk and exploring educational alternatives. *Journal of Adventure Education & Outdoor Learning, 9*(1), 61–77.

Carden, T., Goode, N., & Salmon, P. M. (2017). Not as simple as it looks: Led outdoor activities are complex sociotechnical systems. *Theoretical Issues in Ergonomics Science, 18*(4), 318–337.

Cessford, G. (2009). *National Incident Database report 2007–2008: Outdoor education and recreation.* Retrieved from https://www.incidentreport.org.nz/resources/NID_Report_2007-2008.pdf. Access date: 15th May 2018.

Curtis, R. (1995). *OA guide to outdoor safety management.* 5th September 2016. Retrieved from http://www.princeton.edu/~oa/files/safeman.pdf

Davidson, G. (2007, 20–23 September 2007). *Towards understanding the root causes of outdoor education incidents.* Paper presented at Fifteenth National Outdoor Education Conference, Ballarat, Victoria.

Dekker, S. (2002). Reconstructing human contributions to accidents: The new view on human error and performance. *Journal of Safety Research, 33,* 371–385.

Dickson, T. J. (2012). Learning from injury surveillance and incident analysis. In T. J. Dickson & T. Gray (Eds.), *Risk management in the outdoors: A whole-of-organization approach for education, sport and recreation* (pp. 204–230). Cambridge University Press: Cambridge, GB.

Dickson, T. J., Chapman, J., & Hurrell, M. (2000). Risk in outdoor activities: The perception, the appeal, the reality. *Australian Journal of Outdoor Education, 4*(2), 10–17.

Finch, C. F., & Boufous, S. (2008). Do inadequacies in ICD-10-AM activity coded data lead to underestimates of the population frequency of sports/leisure injuries? *Injury Prevention, 14*(3), 202–204.

Goode, N., Finch, C., Cassell, E., Lenné, M. G., & Salmon, P. M. (2014). What would you like? Identifying the required characteristics of an industry-wide incident reporting and learning system for the led outdoor activity sector. *Australian Journal of Outdoor Education, 17*(2), 2–15.

Gray, S., & Moeller, K. (2010). *Behavioral risk management: Preventing critical incidents in the field.* Paper presented at the Wilderness Risk Management Conference, Colorado Springs, USA.

Haddock, C. (1999). *High potential incidents – Determining their significance: Tools for our trade and a tale or two.* Paper presented at the Proceedings of the 1999 Wilderness Risk Managers Conference, Lander, WY: National Outdoor Leadership School.

Hamonko, M. T., McIntosh, S. E., Schimelpfenig, T., & Leemon, D. (2011). Injuries related to hiking with a pack during National Outdoor Leadership School courses: A risk factor analysis. *Wilderness & Environmental Medicine, 22*(1), 2–6.

Hill, A. (2011). *National Incident Database report 2010: Outdoor education and recreation.* Retrieved from https://www.incidentreport.org.nz/resources/NID_Report _2010.pdf. Access date: 20th June 2018.

Hovelynck, J. (1998). Learning from accident analysis: The dynamics leading up to a rafting accident. *Journal of Experiential Education, 21*(2), 89–95.

Leemon, D. (2009). *Adventure program risk management report: Incident data from 1998–2007.* Retrieved from http://www.aee.org/files/en/user/cms/WRMC _Incident_Poster_text_2008.pdf

Leemon, D., & Schimelpfenig, T. (2003). Wilderness injury, illness, and evacuation: National Outdoor Leadership School's incident profiles, 1999–2002. *Wilderness & Environmental Medicine, 14*(3), 174–182.

Meyer, D. (1979). The management of risk. *Journal of Experiential Education, 2*(2), 9–14.

Mitchell, R., Boufous, S., & Finch, C. F. (1960) & NSW Injury Risk Management Research Centre (2008). *Sport/leisure injuries in New South Wales: Trends in sport/ leisure injury hospitalisations (2003–2005) and the prevalence of non-hospitalised injuries (2005).* Sydney: NSW Injury Risk Management Research Centre, University of New South Wales.

Rasmussen, J. (1997). Risk management in a dynamic society: A modelling problem. *Safety Science, 27*(2/3), 183–213.

Reason, J. (1997). *Managing the risks of organizational accidents.* Burlington, VT: Ashgate.

Salmon, P. M., Williamson, A., Lenné, M., Mitsopoulos-Rubens, E., & Rudin-Brown, C. M. (2009). *The role of human factors in led outdoor activity incidents: Literature review and exploratory analysis.* Retrieved from http://outdoorcouncil.asn .au/wp-content/uploads/2016/08/OAI_REPORT_FINAL_VERSION_OCT _15th_2009.pdf

Salmon, P. M., Williamson, A., Lenné, M., Mitsopoulos-Rubens, E., & Rudin-Brown, C. M. (2010). Systems-based accident analysis in the led outdoor activity domain: Application and evaluation of a risk management framework. *Ergonomics, 53*(8), 927–939. doi:10.1080/00140139.2010.489966

Service Skills Australia. (2010). *National outdoor sector survey 2010: Quantifying the outdoor workforce.* Retrieved from http://www.qorf.org.au/wp-content/uploads/2014/03 /NOSS2010_Report.pdf

Service Skills Australia. (2013). *2013 National outdoor sector survey.* Retrieved from http://qorf.org.au/wp-content/uploads/2014/08/NOSS_2013_Report.pdf

Williams, I., & Allen, N. (2012). *National survey of Australian outdoor youth programs.* Retrieved from http://www.oypra.org.au/resources/OYPRA_Australian_Out door_Survey_Report_2012.pdf

5

Identifying the Needs and Priorities of End Users

Practitioner Summary

Designing an incident reporting system should begin by developing an understanding of the needs and priorities of the end users. This chapter describes a step-by-step process for identifying their needs and priorities, and translating them into design requirements for a new incident reporting system. The process involves gaining consensus from end users on the characteristics of incident reporting systems that they consider to be most important, and then developing design requirements based on these characteristics. The application of the process during the development of UPLOADS is used to illustrate each step. The chapter concludes with a discussion of the design requirements identified for UPLOADS, as most of them are relevant to the design of incident reporting systems in any domain.

Practical Challenges

End users are often concerned with the workload associated with incident reporting. One way to address this concern is restricting the length of the incident report form, however, this needs to be balanced against collecting sufficiently detailed data to support the understanding of incidents from a systems thinking perspective.

5.1 Introduction

While the potential benefits of incident reporting systems are widely acknowledged, their success is highly variable in practice. While there are exemplars of successful incident reporting systems in aviation (Aviation Safety Reporting System, 2016), healthcare (Pharmacopeia, 2008), and rail (Ranney & Raslear, 2012), many incident reporting systems suffer from poor reporting rates and data quality issues (Hill, 2011; Pham, Girard, & Pronovost, 2013; Spigelman & Swan, 2005; Taylor & Lacovara, 2015; Terry et al., 2005). More worryingly, concerns have even been raised that incident reporting systems in healthcare may actually be harmful, as they result in few safety benefits while consuming significant resources (Pham et al., 2013; Pless, 2008). To ensure the uptake and sustainability of an incident reporting system, the needs and priorities of end users need to be identified early in the development process, and incorporated into all aspects of the design.

> Incident reporting systems will fail if they do not address the requirements of the specific context.

This chapter provides a step-by-step guide to identifying the needs and priorities of end users and then translating them into design requirements. The application of the process during the development of UPLOADS is used to practically illustrate each step. The chapter concludes by discussing the design requirements identified for UPLOADS based on the needs and priorities of end users, as most of them are relevant to the design of incident reporting systems in any domain.

5.2 Step-by-Step Guide

The aims of this process (shown in Figure 5.1) are to gain consensus from end users on:

1. Their preferences for the design of supporting materials (i.e. software tools and training materials); and
2. The most important characteristics of an incident reporting system.

This information is then used to collaboratively develop design requirements for the new incident reporting system.

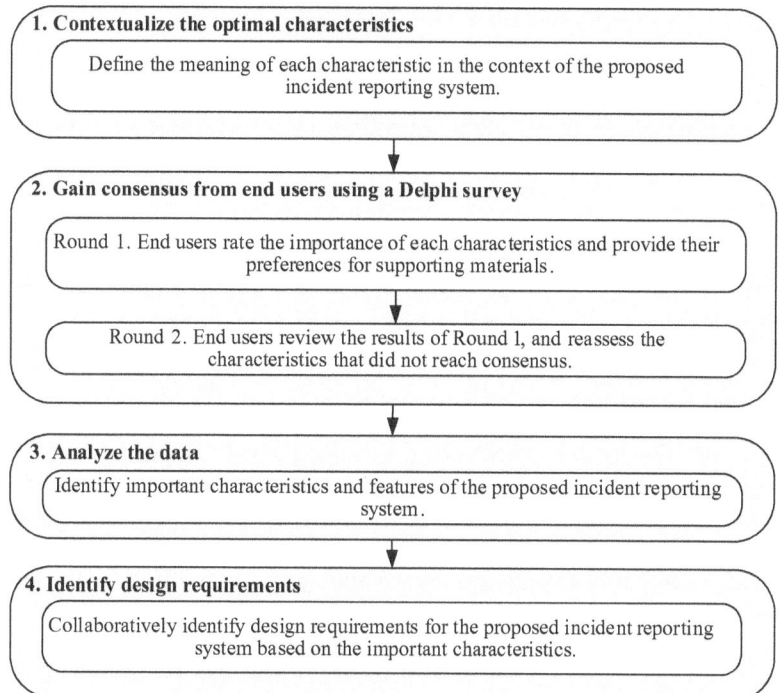

FIGURE 5.1
Step-by-step process for identifying the needs and priorities of end users, and identifying design requirements.

The starting point for this process are the characteristics of an optimal incident reporting system, presented at the end of Chapter 3. These characteristics were identified based on a literature review of good practice for injury surveillance and incident reporting systems across multiple domains (Goode, Finch, Cassell, Lenné, & Salmon, 2014).

While these characteristics are all important in the design of an incident reporting system, some characteristics might not be perceived as particularly important to end users. For example, while the sensitivity and representativeness of the data may be a high priority for people designing an incident reporting, they may be of less importance to end users than the ease of reporting. Similarly, some characteristics may be perceived as highly important in some contexts and not in others. For example, end users contributing data to a sector-wide incident reporting system are likely to be more concerned about data confidentiality, than the end users of an internal incident reporting system. Understanding the priorities of end users at the start of the design process can help ensure that the resulting incident reporting system matches their needs and expectations.

5.2.1 Step 1: Contextualize the Optimal Characteristics

The first step involves defining the meaning of each characteristic in the context of the proposed incident reporting system. This is to ensure that end users understand the implications of rating a characteristic as more or less important.

5.2.1.1 Characteristics Defined for UPLOADS

The definitions developed for each characteristic for UPLOADS are shown in Table 5.1.

5.2.2 Step 2: Gain Consensus from End Users

The second step involves conducting a Delphi survey to obtain consensus on the relative importance of each characteristic, and identifying end user preferences for certain features of the incident reporting system (e.g. format of the training materials, IT platforms).

The Delphi technique is a widely used and accepted method for building a consensus view among subject matter experts (Hasson, Keeney, & McKenna, 2000; Hsu & Sandford, 2007; Linstone & Turoff, 2002). Applying the technique involves asking subject matter experts for their opinions on a specific topic. Multiple surveys are used then to develop a consensus of opinion, in this case the characteristics of the incident reporting system. After each survey, the overall responses of the group are presented back to participants. The findings are usually presented with a summary of the reasons given by participants to support their opinions. Participants can then choose to change their responses in light of the group's opinion, or keep the same response. Provision for feedback requires that there are at least two rounds of surveys (Hsu & Sandford, 2007). There is no pre-determined criteria for consensus, with studies adopting anywhere from 51% to 80% (Keeney, Hasson, & McKenna, 2006). In the context of developing an incident reporting system, it is recommended that a criteria of 80% is used to ensure that the findings represent a near to unanimous view.

The following section describes the Delphi survey undertaken to obtain consensus views from LOA practitioners on the relative importance of the characteristics presented in Table 5.1, and to identify preferences for certain features of the incident reporting system.

5.2.2.1 Delphi Survey for UPLOADS

LOA practitioners who played a key role in managing safety within their organization were invited to participate in the survey.

TABLE 5.1

Optimal Characteristics of Incident Reporting Systems Contextualized for UPLOADS

Characteristic	Definition	Implementation in the Design of UPLOADS
Acceptability	The system matches the characteristics, needs and priorities of the users.	Potential end users are involved in all stages of UPLOADS development to ensure that it meets their needs.
Accessibility	Data are stored in a format that makes it easy to identify relevant cases, and produce aggregate summary reports.	There is an acceptable and reliable turnaround time between information requests from UPLOADS and delivery.
Appropriate use of standardized response options	Standardized response options are utilized to collect data when appropriate.	Classification systems used within UPLOADS are clearly described in a manual that is easily accessible.
Availability	The system is readily available to anyone wishing to report an incident.	UPLOADS should be available to anyone wanting to report an incident or near miss
Clear case definitions	There are clear definitions of what should and should not be reported.	There is a clear definition of what should and shouldn't be entered into UPLOADS.
Clear purpose and objectives	There is a clearly defined reason why the incident reporting system exists and how the data are used.	There is a clearly defined reason why UPLOADS exists and how it is used.
Credible	The data analyses and summaries can be relied upon as accurate reflections of the data.	Data tabulations and summaries of UPLOADS data can be relied upon.
Data collection process described	The method of data collection and the number of steps involved are described in a flowchart.	UPLOADS data collection process and the number of steps involved is clearly documented.
Data completeness	A consistent amount of data is provided about every reported incident.	A consistent amount of data is provided from every case (i.e. each form submitted to UPLOADS database is completely filled out).
Data confidentiality and individual privacy	Information that could be used to identify individuals or organizations is not revealed in any outputs or data summaries.	Neither specific individuals nor particular organizations are identifiable in outputs and data summaries generated from UPLOADS.
Ease of reporting	Submitting an incident report is as easy as possible.	Submitting a report to UPLOADS is as easy as possible.

(Continued)

TABLE 5.1 (CONTINUED)

Optimal Characteristics of Incident Reporting Systems Contextualized for UPLOADS

Characteristic	Definition	Implementation in the Design of UPLOADS
Expert analysis	Reports are validated and evaluated by experts who understand the circumstances and are trained to analyze the contributory factors involved in the incident.	The data stored within UPLOADS is interpreted by experts.
Flexibility	The system is easy to change, especially when evaluation shows that change is necessary or desirable.	UPLOADS can adapt to meet new requirements (e.g. a new led outdoor activity can be added to the database readily).
Comprehensive guidance	Training materials support the appropriate reporting, analysis, and interpretation of incident data.	UPLOADS reports clearly explain the results and what they mean.
Non–punitive	Reporters are free from the fear of retaliation or punishment as the result of reporting.	Reporters of incidents or near misses are free from the fear of retaliation or punishment as the result of reporting to UPLOADS.
Positive predictive value	The incident report provides an accurate description of the incident.	The data entered into UPLOADS is an accurate description of the incident.
Quality control measures	Data validation is used to ensure that data are entered into the tool accurately.	The quality of the data entered into UPLOADS is monitored and maintained by researchers.
Representativeness	The data accurately represents how frequently incidents, and the contributory factors involved, are occurring over time.	Results from UPLOADS accurately represent how frequently incidents and near misses are occurring during led outdoor activities.
Responsive	End users are committed to changing the work system based on the findings.	Stakeholders are committed to changing practices based on recommendations from UPLOADS.
Sensitivity	All relevant incidents that occur are reported.	All relevant cases are entered into UPLOADS (i.e. all incidents that occur at the organizations involved in the project are entered into the database).

(Continued)

TABLE 5.1 (CONTINUED)

Optimal Characteristics of Incident Reporting Systems Contextualized
for UPLOADS

Characteristic	Definition	Implementation in the Design of UPLOADS
Simplicity	The database for storing information has a simple structure and is easy to operate.	UPLOADS has a simple structure and is easy to operate.
Specificity	No irrelevant cases are reported.	No irrelevant cases are entered into UPLOADS.
Stability of the system	The system is reliably available for data input at all times.	UPLOADS should be reliably available for data input at all times.
Sustainability of the system	The system is easy to maintain and update.	Participating organizations should be able to easily maintain and update UPLOADS.
Sustained leadership support	Senior leaders are committed to learning from incidents on an ongoing basis.	Organizations involved in the project should be committed to contributing data on an ongoing basis.
System security	Data access is controlled to safeguard against disclosure of confidential material.	The data entered into UPLOADS is password protected.
Systems-oriented	Incident prevention strategies focus on redesigning the work system to support safe operation.	UPLOADS recommendations focus on changes that could or should be made to policies, procedures or activities, rather than being targeted at activity leaders.
Timeliness	There is rapid turnaround between data collection and reporting meaningful information to stakeholders.	There is rapid turnaround between data collection and reporting meaningful information to stakeholders.
Usefulness	The scheme includes sufficiently detailed classifications so that aggregate analyses of incidents can easily be interpreted.	The data summarized from UPLOADS can be used to identify ways to reduce incidents and near misses.
Utility	The tool is practical and affordable, and does not put unnecessary workload on reporters or organizations.	UPLOADS data collection process is practical and does not place an undue burden on participating organizations or the people who contribute to it.

Source: From Goode, N., Finch, C., Cassell, E., Lenné, M. G., & Salmon, P. M. 2014. *Australian Journal of Outdoor Education*, 17(2), 2–15. With permission.

Twenty-five people agreed to participate from organizations representing a diverse cross-section of the outdoor sector, including school-based groups (n = 4), government organizations (n = 7), faith-based organizations (n = 5), not-for-profit organizations (n = 6), and commercial providers (n = 3).

Participating involved completing two rounds of online surveys, in which respondents were asked to rate the importance of the characteristics presented in Table 5.1. Consensus was defined as at least 80% agreement.

The first survey was used to gather information on preferences on the reporting criteria, training, and data collection methods, as well as initial opinions on the importance of the characteristics presented in Table 5.1. LOA practitioners were asked to rate whether each characteristic was:

- *Essential* (i.e. UPLOADS must have this characteristic);
- *Ideal* (i.e. in an ideal world where money, time, and other resources are unlimited, it would be good if UPLOADS had this characteristic); or
- *Not required* (i.e. this characteristic would be of no value or use for UPLOADS).

In the second survey, the LOA practitioners were presented with a summary of the ratings of each characteristic from the first survey, and asked to reassess the characteristics that did not reach consensus as either:

- *Yes* (UPLOADS must have this characteristic); or
- *No* (UPLOADS does not require this characteristic).

This gave the LOA practitioners an opportunity to revise their original ratings, and to more categorically state whether they thought the remaining characteristics were required.

5.2.3 Step 3: Analyze the Data

The next step involves analyzing the data that has been collected to identify the required characteristics and features of the proposed incident reporting system. This involves calculating the percentage agreement for each characteristic and determining whether each characteristic meets the criteria for consensus (i.e. >80% agreement). For the other features, the percentage of participants who indicated that they would like each feature should be calculated.

The following sections presents a summary of findings from the two rounds of the Delphi survey conducted for UPLOADS.

5.2.3.1 Analysis of Delphi Survey for UPLOADS

A summary of the LOA practitioners' ratings of the potential features of UPLOADS is shown in Table 5.2. The findings suggested that the reporting criteria for UPLOADS should be reasonably broad, and not just focus on injuries. Participants indicated a preference for training materials presented across multiple modalities, rather than identifying a preference for any format (e.g. video, manual, online). Similarly, there was a desire for both paper forms and a smart phone application for reporting incidents.

A summary of the consensus that was reached on the importance of the characteristics for UPLOADS across the two surveys is shown in Table 5.3. Out of the 30 characteristics, there was a consensus that 13 were essential, and 13 were required.

5.2.4 Step 4: Identify Design Requirements

The final step in the process is identifying design requirements for the proposed incident reporting system based on the characteristics considered essential or required by end users. The design requirements should be developed in collaboration with end users (i.e. in a workshop) to ensure that they meet their needs and expectations.

The following section describes the design requirements that were identified for UPLOADS during a workshop with our industry partners. Most of these requirements are relevant to the design of incident reporting systems in any domain.

5.2.4.1 Design Requirements for UPLOADS

The design requirements that were identified based on the 'essential' characteristics are shown in Table 5.4. The 'essential' characteristics highlighted end-user concerns with the workload associated with incident reporting

TABLE 5.2

Summary of LOA Practitioners' Ratings of Desirable Features of UPLOADS Supporting Materials – Survey 1

Features	Consensus on Features
Criteria for reporting (types of incidents)	Injuries (96%); Near misses (76%); Property damage (76%) Additional: Behavioral or psychological outcomes, environmental hazards
Criteria for reporting (injury severity)	All injuries (56%); Only more serious injuries (36%); Unsure (8%)
Training	Online video tutorial (76%); Online manual (72%); Hands-on seminar (52%); Paper-based manual: 24%
Data collection methods	Paper-based forms (84%); Smart phone application (88%)

TABLE 5.3

Characteristics Identified through Consensus as Essential
or Required for UPLOADS

Characteristic	80% Agreement
Clear case definitions	Essential
Credible	Essential
Data confidentiality and individual privacy	Essential
Ease of reporting	Essential
Guidance material for data interpretation	Essential
Non-punitive	Essential
Positive predictive value	Essential
Representativeness	Essential
Simplicity	Essential
Use of uniform classification systems	Essential
Usefulness	Essential
Utility	Essential
Sustainability of the system	Essential
Accessibility	Required
Availability	Required
Clear purpose and objectives	Required
Data collection process described	Required
Data completeness	Required
Flexibility	Required
Quality control measures	Required
Sensitivity	Required
Specificity	Required
Stability of the system	Required
Sustained leadership support	Required
System security	Required
Systems-oriented	Required

(e.g. ease of reporting, utility, simplicity), confidentiality (e.g. non-punitive, individual privacy), and the trustworthiness of the data (e.g. credibility, clear case definitions, use of uniform classification systems, and representativeness). To address the concerns around workload, our industry partners determined that the initial incident report should be limited to two pages. In addition, it was determined that the design of the incident reporting system should focus on minimizing the administrative duties associated with data entry, analysis and submitting data to the National Incident Dataset.

The concerns regarding confidentiality were interpreted in terms of the legal ramifications of contributing potential sensitive incident data to a sector-wide incident reporting system. Several design requirements were identified to ensure anonymity of reporting for organizations and privacy for individuals involved in incidents. In particular, UPLOADS was designed

TABLE 5.4

Design Requirements to Operationalize Characteristics Considered as 'Essential'

Characteristic	Design Requirements
Utility	Length of incident report form limited to two A4 pages
	The system minimises double entry of information
Usefulness	Tools for analysing the data, as well as collecting it
	Training material on how to develop appropriate countermeasures from findings
	Committee to develop appropriate, sector-wide initiatives from findings
Ease of reporting	Staff can access reporting forms at any time
Clear case definitions	Incident severity scale to clearly define the type of incidents to be reported
	Clear definitions of 'incident', 'adverse outcome', and 'near miss'
Representativeness	System to support the collection of participation data in addition to incident data
	Data submitted to the sector-wide database in deidentified format, to discourage 'deletion' of sensitive cases
Positive predictive value	Each incident report reviewed by manager within the organization
	Where appropriate, all the people involved in the incident to contribute to and review incident reports
Non-punitive	Individual reports cannot be linked back to organizations within the sector-wide database
	Training material for organizations on how a 'just culture' supports reporting
Credible	Sector-wide aggregate data to be analyzed by a researcher
	Sector-wide reports undergo a peer-review process before release
Use of uniform classification systems	Domain-specific classification schemes to code the contributory factors involved in incidents
	International Classification of Diseases, 10th edition (World Health Organization, 1994) codes used to classify injury types
	Incident severity scale to classify incidents
Sustainability (system)	UPLOADS should not require any special computer skills to operate it
	Sufficient training materials provided so that operation of the system can be handed over if staff member leaves
Simplicity	The prototype UPLOADS will be reviewed by multiple stakeholders to ensure it is intuitive and domain-appropriate
Data confidentiality and individual privacy	UPLOADS designed as a decentralized reporting system, where organizations collect and manage their own data
	Sector-wide data reported at the aggregate level only
	Limits to be set around the reporting of incident types (e.g. more than three kayaking incidents required to report on this type of incident)
Guidance material	Reports on the sector-wide data to be written in lay language
	Guidance on interpreting complex analyses provided in lay language

Source: Adapted from Goode, N., Finch, C., Cassell, E., Lenné, M. G., & Salmon, P. M. 2014. *Australian Journal of Outdoor Education*, 17(2), 2–15. With permission.

TABLE 5.5

Design Requirements for Operationalizing Characteristics Considered as 'Required'

Characteristic	Strategies
Clear purpose and objectives	There is a clearly defined reason why UPLOADS exists
	The type of outdoor organizations that UPLOADS is intended to service clearly defined
	The type of activities that UPLOADS collects data on clearly defined
	The potential uses of the data will be clearly defined
Data collection process described	A flow diagram to illustrate the steps involved in the data collection process
Accessibility	Clear dates set for the release of sector-wide reports
	Requests for specific data analyses to be negotiated on a case basis
Flexibility	Additional fields can be added to the database to meet the needs of individual organizations
	Easy to update the UPLOADS software
Stability of the system	The UPLOADS software will run on PCs within each contributing organization to avoid potential issues with a central sever
	The UPLOADS software will not rely on a connection to the Internet
Quality control measures	Researchers will conduct periodic studies to assess the reliability and validity of the coding taxonomy
	Researchers will conduct periodic checks to ensure the database fields are being used correctly
Systems-orientated	UPLOADS will be based on a widely accepted systems-orientated accident causation model
Sustained leadership support	Organizations commit to contributing data for specified periods of time (e.g. 1 year), after which they may choose to opt out
	Peak bodies and professional associations promote UPLOADS as best practice within the sector
System security	Organizations manage their own data, and submit deidentified data to UPLOADS on a periodic basis
	UPLOADS software is password protected
Data completeness	Mandatory fields are used to specify the minimum requirement for an incident report
Sensitivity	Data submitted to the sector-wide database in deidentified format, to discourage 'deletion' of sensitive cases
	Incident severity scale to clearly specify the types of incidents to be reported
Availability	Material provided to train all staff within the organization on how to report an incident
Specificity	Incident severity scale to clearly specify the types of incidents to be reported
	Cases in the sector-wide database reviewed by researchers to ensure they meet the criteria for an incident or near miss

Source: Adapted from Goode, N., Finch, C., Cassell, E., Lenné, M. G., & Salmon, P. M. 2014. *Australian Journal of Outdoor Education*, 17(2), 2–15. With permission.

as a decentralized, sector-wide incident reporting system, where organizations collected and managed their own incident data. In addition, the incident reporting software was designed to automatically remove all information identifying individuals and organizations prior to submission to the sector-wide database. A response protocol for legal requests for information was also set-up, which explained that it is not possible to identify individual cases or organizations from the dataset. Finally, several constraints on reporting findings from the sector-wide database were specified so that the reports did not unintentionally identify organizations contributing incident data.

The final concern around the trustworthiness of the data was addressed in several ways. Validated classification schemes were used to ensure that the incident data were collected in an appropriate and structured way. The incident report form was also designed to collect detailed qualitative information (i.e. using free text fields with prompts) to ensure that responses to standardized response options could be verified.

Table 5.5 outlines the design requirements that were identified based on the characteristics considered to be 'required' within UPLOADS. Again, the 'required' characteristics confirmed that the trustworthiness of the data was a high priority for the LOA sector. In addition, they also highlighted some concerns around the system infrastructure and security. It should be noted that while Table 5.5 outlines the design requirements that were used to address these concerns during the design of UPLOADS, technology has moved on significantly since this time (e.g. Internet stability and the storage of data). This means that the design requirements identified to address issues with the stability of the system (e.g. running the software on PCs within each contributing organization) are not relevant to the design of future incident reporting systems. The most recent version of UPLOADS (2018) now uses cloud-based technology, which was not readily available during the initial development.

5.3 Summary and Next Steps

This chapter described a step-by-step process for identifying the needs and priorities of end users for the design of a proposed incident reporting system. The process involves adapting the definitions of the characteristics of an optimal incident reporting system for the context, gaining consensus from end users over the relative importance of each characteristic, and then identifying design requirements based on the important characteristics. The application of the process during the development of UPLOADS was used to illustrate each step. For readers wishing to develop an incident reporting system in their own context, it is recommended that a similar process is adopted. This provides the basis for developing a tool that is acceptable to end users.

Further Reading

A detailed description of the procedure and materials used to conduct the Delphi survey is presented in this paper: Goode, N., Finch, C., Cassell, E., Lenné, M. G., & Salmon, P. M. (2014). What would you like? Identifying the required characteristics of an industry-wide incident reporting and learning system for the led outdoor activity sector. *Australian Journal of Outdoor Education, 17*(2), 2–15.

References

Aviation Safety Reporting System. (2016). *ASRS Program Briefing.* 3rd December 2017 Retrieved from https://asrs.arc.nasa.gov/docs/ASRS_ProgramBriefing2016.pdf

Goode, N., Finch, C., Cassell, E., Lenné, M. G., & Salmon, P. M. (2014). What would you like? Identifying the required characteristics of an industry-wide incident reporting and learning system for the led outdoor activity sector. *Australian Journal of Outdoor Education, 17*(2), 2–15.

Hasson, F., Keeney, S., & McKenna, H. (2000). Research guidelines for the Delphi survey technique. *Journal of Advanced Nursing, 32*(4), 1008–1015.

Hill, A. (2011). *National Incident Database Report 2010: Outdoor Education And Recreation.* Retrieved from https://www.incidentreport.org.nz/resources/NID _Report_2010.pdf

Hsu, C.-C., & Sandford, B. A. (2007). The Delphi technique: Making sense of consensus. *Practical Assessment, Research & Evaluation, 12*(10), 1–8.

Keeney, S., Hasson, F., & McKenna, H. (2006). Consulting the oracle: Ten lessons from using the Delphi technique in nursing research. *Journal of Advanced Nursing, 53*(2), 205–212.

Linstone, H. A., & Turoff, M. (2002). *The Delphi method: Techniques and applications* (Vol. 18): Addison-Wesley Publishing Company, Advanced Book Program.

Pham, J. C., Girard, T., & Pronovost, P. J. (2013). What to do with healthcare incident reporting systems. *Journal of Public Health Research, 2*(3).

Pharmacopeia, U. (2008). 8th annual MEDMARX® report indicates look-alike/ sound-alike drugs lead to thousands of medication errors nationwide. *US Pharmacopeia.*

Pless, B. (2008). Surveillance alone is not the answer. *Injury Prevention, 14*(4), 220–222. doi:10.1136/ip.2008.019273

Ranney, J., & Raslear, T. (2012). Derailments Decrease at a C3RS Site at Midterm. U.S. DOT Federal Railroad Administration RR12-04. Retrieved from http://www .fra.dot.gov/eLib/details/L03582 (2012, April).

Spigelman, A. D., & Swan, J. (2005). Review of the Australian incident monitoring system. *ANZ Journal of Surgery, 75*(8), 657–661.

Taylor, J. A., & Lacovara, A. V. (2015). From infancy to adolescence: The development and future of the national firefighter near-miss reporting system. *New Solutions: A Journal of Environmental and Occupational Health Policy, 24*(4), 555–576. doi:10.2190/NS.24.4.h

Terry, A., Mottram, C., Round, J., Firman, E., Step, J., & Bourne, J. (2005). *A safer place for patients: Learning to improve patient safety.* Technical Report. The Stationary Office, London. http://eprints.whiterose.ac.uk/3427/1/NAO_2005_a_safer_place .pdf. Access date: 15th May 2018.

World Health Organization. (1994). *International Classification of Diseases, 10th edition (ICD-10).* Retrieved from http://apps.who.int/classifications/icd10/browse /2010/en. Access date: 15th May 2018.

6

Adapting Accimap for Use in an Incident Reporting System

Practitioner Summary

This chapter describes how to adapt Accimap for use within an incident reporting system. The process involves two tasks: (1) adapting the labels on the levels of the Accimap framework to better describe the structure of the system in question; and (2) developing a domain-specific classification scheme describing the contributory factors found at each level of the Accimap framework. The chapter describes how to undertake these tasks by drawing on multiple data sources, including workshops with subject matter experts, the literature, case studies of fatal incidents, and incident reports. The adaptation of Accimap for use within UPLOADS is used to illustrate the process.

Practical Challenges

Adapting Accimap for use within an incident reporting system requires the development of a classification scheme which comprehensively describes the contributory factors at each level of the Accimap framework. It is usually impossible to develop a comprehensive scheme based soley on existing incident reports, as they rarely include any contributing factors at the higher levels of the framework. To address this problem, we recommend that additional data sources are utilized, such as subject matter expert input, the literature, and case studies of major incidents and fatalities.

6.1 Introduction

As discussed in Chapter 2, Accimap is a usable and readily adaptable analysis method that has consistently shown its explanatory power for understanding incident causation in a range of domains (Salmon, Walker, Read, Goode, & Stanton, 2016; Waterson, Jenkins, Salmon, & Underwood, 2016). One limitation of Accimap, however, is that it does not provide a classification scheme to guide analysts in identifying and classifying contributory factors. Rather, analysts are required to identify contributory factors and relationships from the data, and then develop descriptive labels for representing them on the Accimap framework. This creates concerns regarding the reliability and validity of the method. It also makes it difficult to aggregate analyses across multiple incidents, and prevents the implementation of Accimap within incident reporting systems (Goode, Salmon, Taylor, Lenné, & Finch, 2017; Salmon, Cornelissen, & Trotter, 2012).

This chapter provides guidance on adapting Accimap to support the consistent analysis of incidents and classification of contributory factors, as required as part of an incident reporting system. The processes used to adapt Accimap for UPLOADS are used to illustrate each step.

6.2 Tasks Involved in Adapting Accimap

There are two tasks involved in adapting Accimap for use in an incident reporting system:

1. Adapting the labels on the levels of the Accimap framework to better describe the structure of the system in question (e.g. healthcare, rail, manufacturing); and

2. Developing a domain-specific classification scheme describing the contributory factors found at each level of the Accimap framework.

Guidance on undertaking these tasks is provided in the following sections.

6.3 Adapting the Levels on the Accimap Framework

Accimap represents the system under analysis as a hierarchy comprised of multiple levels. As discussed in Chapter 3, the levels are often adapted to reflect the structure of the system under analysis. Table 6.1 presents examples from public health, policing, manufacturing and road freight transport. Adapting the levels involves three steps.

TABLE 6.1

Examples of Adaptations of the Labels on the Levels of Accimap to Describe Different Systems

Rasmussen and Svedung (2000)	Waterson (2009)	Jenkins, Salmon, Stanton, and Walker (2010)	Le Coze (2010)	Newnam and Goode (2015)
Generic	*Public Health (Hospitals)*	*Policing*	*Manufacturing Society, Market*	*Road Freight Transport*
Government Policy and Budgeting	Government	Government Policy and Budgeting	Government, Regulatory System	Government Bodies
Regulatory Bodies and Associations	Regulatory	Regulatory Bodies and Associations	–	Regulatory Bodies
Local Area Government, Company Management, Planning and Budgeting	Trust Governance	Strategic Command	Company	Other Organizations and Clients
Technical and Operational Management	Hospital Management	Tactical Command	Site Management	Heavy Vehicle Companies
Physical Processes and Actor Activities	Clinical Management	Physical Processes and Actor Activities	Operational Management	Drivers and Other Actors at the Scene of the Incident
Equipment and Surroundings	Equipment and Surroundings	Equipment and Surroundings	Shop Floor and Installations	Equipment, Surroundings, and Meteorological Conditions

6.3.1 Step 1: Construct an Actormap of the System under Analysis

The first step involves mapping out the people, technologies, and organizations involved in managing safety in the domain onto the original Accimap framework. Rasmussen (1997) refers to this as an Actormap.

An Actormap can be constructed based on documentation (e.g. regulations, incident and investigation reports), and then refined with end users during a workshop or meeting. It is also useful to consider whether the *number* of levels accurately reflect the structure of the system under analysis. For example, in his analysis of a manufacturing incident, Le Coze (2010) combined the government and regulatory levels of the framework, and added a level describing the influence of society and the market.

6.3.2 Step 2: Relabel the Levels

The next step is relabeling the levels to more accurately reflect the people, technologies, and organizations represented at each level of the Actormap. This should be undertaken collaboratively with end users to ensure that they understand the labels given to the levels.

6.3.3 Step 3: Test and Refine the Framework

The final step is testing and refining the framework based on existing incident data. This involves classifying the contributory factors identified from incident data using the levels of the adapted framework, and testing whether the labels clearly distinguish between the types of factors that should, and should not, be classified at each level.

6.3.4 Adapting the Accimap Framework for UPLOADS

For UPLOADS, the system under analysis is the LOA system in Australia. To adapt the Accimap levels to describe this system, a workshop was first held with LOA subject matter experts. The experts were asked to describe the LOA system in terms of the activity context, the key people involved in LOAs, and the people and agencies that impact on the planning and delivery of LOAs (see Figure 6.1).

The people and organizations identified by the experts were then represented on the original Accimap framework. The experts then renamed each level to better describe the LOA system, and the types of contributory factors that would be identified during an incident analysis. For example, 'physical processes and actor activities' was changed to 'decisions and actions of leaders, participants and other actors at the scene of the incident'.

Following the workshop, the labels given to each level were then refined and validated based on contributory factors identified from Australian and New Zealand incident data. The adapted Accimap framework (shown in Figure 6.1) was then used as a starting point for the development of the UPLOADS contributory factor classification scheme. The Actormap was also useful for developing the process for learning from incidents for UPLOADS, as it identified the stakeholders who share the responsibility for the safe delivery of LOAs in Australia, and need to be involved in designing and implementing incident prevention strategies.

6.4 Developing a Contributory Factor Classification Scheme

A contributory factor classification scheme is a taxonomy for describing the different types of contributory factors involved in incidents, based on the underpinning model of accident causation. For example, HFACS (shown in Table 6.2) includes a four level framework based on Reason's (1990) Swiss Cheese model of accident causation with categories describing the contributory factors at each level framework. The idea is that analysts can use the HFACS categories to classify contributory factors identified from incident reports. This makes it easier to summarize the findings and identify recurring contributory factors across multiple incidents.

In developing a new contributory factor classification scheme, it is good practice to start by developing a prototype based on existing incident data. Prior to implementation in an incident reporting system, the prototype scheme should then be tested and refined until satisfactory reliability and validity are achieved. As discussed in Chapter 3, reliability and validity are important requirements because summaries of the findings will provide a misleading picture of incident causation if analysts cannot accurately and consistently apply the scheme. Developing a scheme that is reliable and valid usually requires multiple rounds of testing and revision following the initial development of the prototype (see Chapter 7 for further guidance).

The following section presents some general principles that can help ensure that a prototype classification scheme ultimately achieves satisfactory reliability and validity. The process required to develop a prototype scheme is then described.

6.4.1 Principles for Designing a Prototype Classification Scheme

There are four principles that should underpin the initial design of a prototype contributory factor classification scheme:

1. The categories should be mutually exclusive;
2. The categories should be exhaustive;

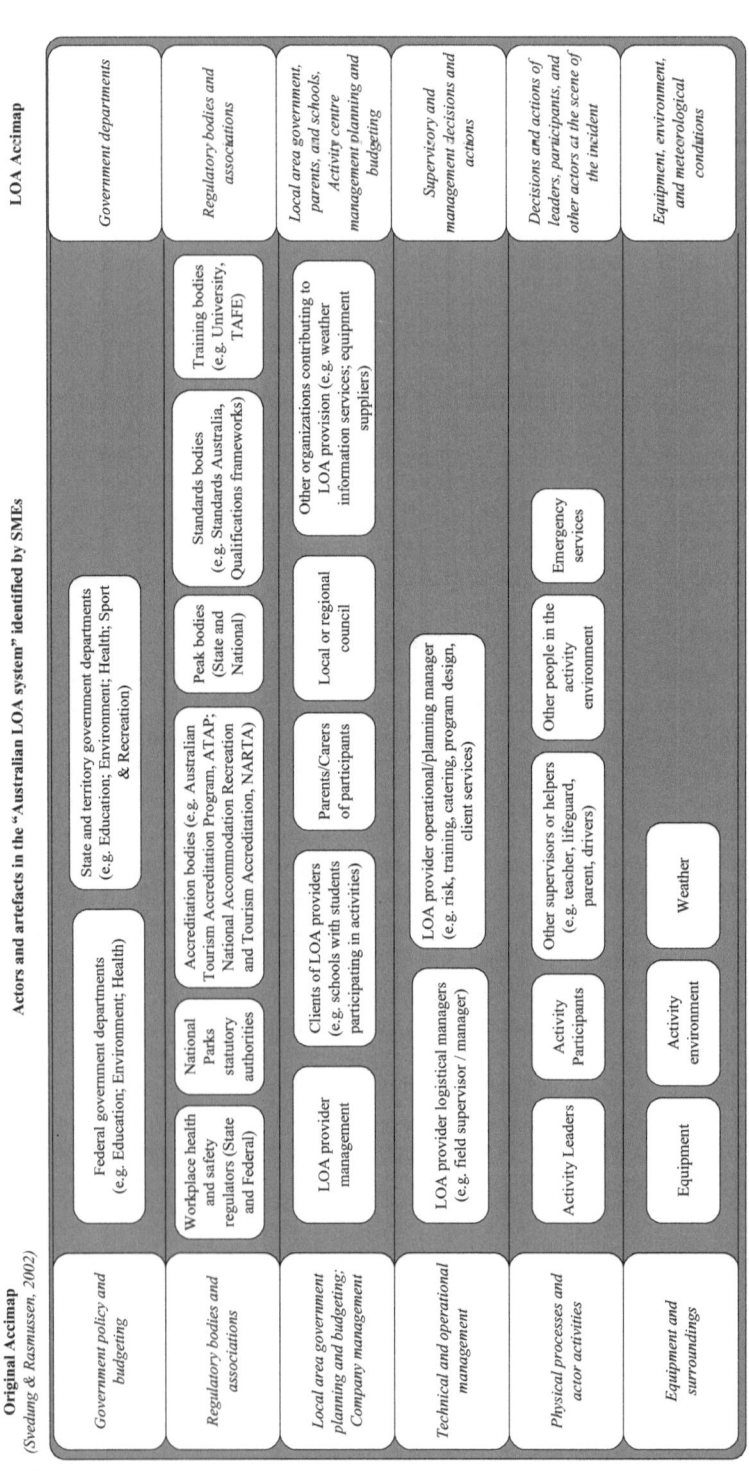

FIGURE 6.1

The original Accimap framework, the human and non-human agents identified at each level of the framework by experts, and the final validated framework representing the LOA system.

TABLE 6.2

The HFACS Four Level Framework and Associated Contributory Factor Categories

Level	Categories
UNSAFE ACTS	1. Errors 1.1. Decision Errors 1.2. Skill-Based Errors 1.3. Perceptual Errors 2. Violations 2.1. Routine 2.2. Exceptional
PRECONDITION FOR UNSAFE ACTS	3. Environmental Factors 3.1. Physical Environment 3.2. Technological Environment 4. Condition of Operators 4.1. Adverse Mental State 4.2. Adverse Physiological State 4.3. Physical/Mental Limitations 5. Personal Factors 5.1. Crew Resource Management 5.2. Personal Readiness
UNSAFE SUPERVISIONS	6. Inadequate Supervision 7. Planned Inappropriate Operations 8. Failed to Correct Problem 9. Supervisory Violations
ORGANIZATIONAL INFLUENCES	10. Resource Management 11. Organizational Climate 12. Organizational Process

Source: Adapted from Wiegmann, D. A., & S. A. Shappell 2003. *A human error approach to aviation accident analysis: The human factors analysis and classification system.* Burlington, VT: Ashgate.

3. The scheme should be arranged in a hierarchical structure; and

4. The categories should be labeled neutrally, rather than negatively.

Mutually exclusive categories means that there is little to no overlap between the types of contributory factors assigned to one category within the scheme, and those assigned to another category. This helps analysts to clearly differentiate between the categories, and reliably assign contributory factors to the correct category. For example, it would be difficult for analysts to differentiate between the categories: '*inappropriate* equipment for the task' and '*wrong* equipment for the task'. To reduce any perceived overlap, clear definitions of what should, and should not be, classified under each code should be specified (Wallace & Ross, 2006b).

Exhaustive categories means that the classification scheme describes all of the possible contributory factors at each level of the Accimap framework. If the available evidence on incident causation does not adequately describe contributory factors at a particular level of the Accimap framework (i.e. Government, Regulation), then additional categories should be developed based on domain expert knowledge. Expert knowledge should also be used to initially check whether the codes at each level of the framework appear to be comprehensive.

In general, categories labeled 'other' should be avoided. Novice analysts tend to assign all contributory factors they are uncertain about to the 'other' category, regardless of whether there is a more appropriate category available (Wallace & Ross, 2006b). This obviously reduces the usefulness of any summaries of the data, as well as impacting on reliability and validity.

It is, however, quite likely that new and unexpected contributory factors will emerge as new incident data is collected, especially if the incident reporting system is designed to collect comprehensive data on incident causation from a systems thinking perspective. As a compromise, a single 'other' category should be included to capture previously unknown contributory factors. If the 'other' category needs to be used regularly during incident analysis, this provides an indicator that the scheme needs to be updated.

Classification schemes with *hierarchical structures* make it easier for novice users to understand the categories, and the relationships between them. Again, this improves the likelihood that the scheme will produce accurate and reproducible results (Wallace & Ross, 2006b). For example, Level 1 of HFACS includes two categories: 'Errors' and 'Violations'. The category 'Errors' is further divided into the sub-categories 'Decision Errors', 'Skill-Based Errors', and 'Perceptual Errors'.

A hierarchical structure also makes it easier to analyze incident reports with different levels of detail. Using the example of HFACS again, an incident report describing a 'pilot error' would be classified under the higher-level category 'Error'. If the incident report further specifies 'pilot error due to the application of the inappropriate procedure', this would be classified under the Error sub-category 'Decision Error'. However, if no further details were provided, just choosing 'Error' would be appropriate. The hierarchical structure makes HFACS more flexible, and easier to apply across different incident reports.

One caveat is that a hierarchical structure with too many levels may lead to *poor* reliability because it is very difficult to develop *highly specific* and *mutually exclusive* codes (Gordon, Flin, & Mearns, 2005; O'Connor, 2008; O'Connor, Walliser, & Philips, 2010; Olsen & Shorrock, 2010). For example, the Department of Defense Human Factors Analysis and Classification System (DOD-HFACS) has the same four-level structure of the original HFACS (Table 6.2), plus another layer of more detailed categories. For example, 'Skill-Based Errors' are further sub-divided into: 'Inadvertent Operation', 'Checklist Error', 'Procedural Error', 'Overcontrol/Undercontrol', 'Breakdown in Visual Scan', and 'Inadequate Anti-Government Straining Maneuver'. It is easy to see how a contributory factor might be categorized as a Checklist Error and Procedural error. So, while detailed categories are desirable to enhance the usefulness of the coded data, this needs to be carefully balanced against ensuring that they are also mutually exclusive.

Finally, categories should be labeled neutrally, rather than negatively, to make them more flexible. For example, a neutral category label might be 'Equipment', while a negative category label might be 'Lack of equipment'. The neutrally labeled category can be used to classify a 'lack of equipment', as well as other problems relating to the availability and suitability of the equipment. The negatively labeled category only applies to one specific problem. Neutral labels are also desirable from a systems thinking perspective because the contributory factors underpinning incidents are not necessarily errors or failures, or violations – they may just be the normal conditions of work. Neutrally labeled categories are capable of classifying both types of contributory factors.

6.4.2 Developing a Prototype Contributory Factor Classification Scheme

Developing a prototype contributory factor classification scheme for Accimap involves:

1. Identifying multiple sources of data describing incident causation in the domain;
2. Thematically analyzing each data source to identify different types of contributory factors, and classifying them according to the levels of the adapted Accimap framework; and
3. Integrating the contributory factor categories identified from the different data sources to create the scheme (in line with the principles described in Section 6.4.1).

Depending on the quality of the data sources, this process should result in a highly detailed prototype classification scheme, with categories describing contributory factors at each level of the Accimap framework. The application of the process during the development of UPLOADS is described in the following section.

6.4.3 UPLOADS Prototype Contributory Factor Classification Scheme

Four data sources were utilized to develop the prototype UPLOADS contributory factor classification scheme:

1. Literature on the contributory factors involved in LOA incidents;
2. Case study analyses of fatal LOA incidents;
3. Incident reports from Australian LOA providers; and
4. Incident reports from the New Zealand National Incident Database (NZ NID).

These data sources were thematically analyzed to identify different types of contributory factors. The thematic analysis of each data source involved:

1. Extracting the contributory factors from each source and creating a list of categories;
2. Grouping similar categories from different sources together and creating a hierarchical structure; and
3. Classifying the final categories according to the levels of the adapted Accimap framework (Figure 6.1).

The final categories were then reviewed by a second analyst and end users to ensure that the labels given to the categories, and the hierarchical structure, were easy to understand. The categories were then revised based on this feedback, and again mapped onto the adapted Accimap framework.

The following sections describe the data sources in more detail, and the contributory factors identified from each analysis. This is intended to provide an example for readers wishing to develop their own classification scheme. The prototype classification scheme is then presented.

6.4.3.1 Literature on the Contributory Factors Involved in LOA Incidents

A review was undertaken to identify literature describing the contributory factors involved in LOA incidents. The final selection of documents included: analyses of LOA incidents, LOA accident causation models, and general discussions of the role that specific contributory factors play in incident causation. A summary of the types of contributory factors that were identified from in the literature is presented in Figure 6.2.

6.4.3.2 Case Study Analyses of Fatal LOA Incidents

Two fatal incidents were selected to inform the development of the classification scheme: the Lyme Bay sea canoeing incident; and the Mangatepopo Gorge walking incident. These incidents were selected based on the quality and depth of the available information, as both incidents were described in detail in publicly available inquiry and investigation reports.

The Lyme Bay case study analysis was based on the official inquiry report into the incident (Jenkins & Jenkinson, 1993); it provided an exhaustive account of the incident and the contributory factors. This incident resulted in the death of four students whilst on an outdoor education program on the 22nd March 1993. The program involved an introductory open sea canoeing activity in Lyme Bay, Dorset, UK, with a group of eight students, a school teacher, a junior instructor and a senior instructor. After a series of capsizes, the junior instructor and students were separated from the group and blown out to sea. The junior instructor and

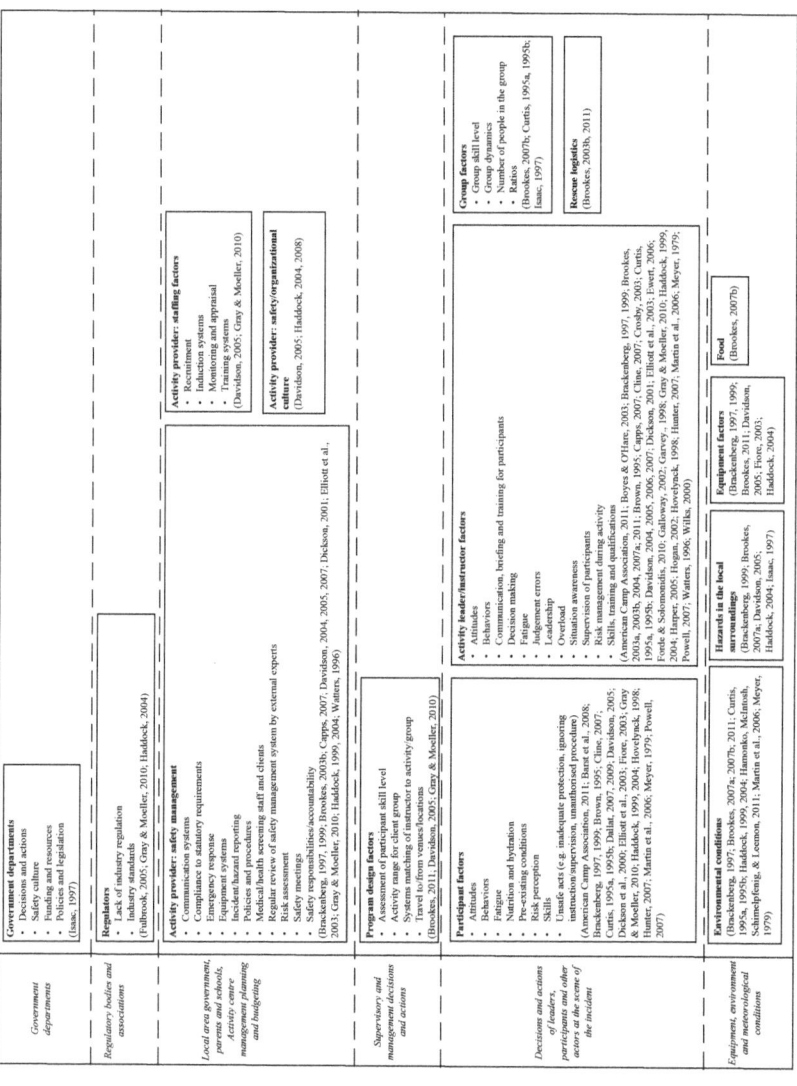

FIGURE 6.2
Summary of the types of contributory factor identified in the literature review, with the relevant references (see Appendix A for the reference list).

students were subsequently swamped, and their canoes washed away, by high winds and waves. Although the group had been due back at noon, emergency services were not called until 15:30. Four students drowned as a result, while the others were rescued by helicopter. The contributory factors, and the relationships between them, identified from the analysis of this incident is shown in Figure 6.3.

The Mangatepopo Gorge case study analysis was based on the coroner's inquiry (Davenport, 2010) and independent investigation (Brookes, Smith, & Corkill, 2009) reports on the incident. This incident has been described in detail in Chapter 2. The contributory factors and the relationships between them identified from the analysis of this incident are shown in Figure 6.4.

6.4.3.3 Australian and New Zealand Incident Reports

As well as reviewing fatal incidents, an important part of developing the classification scheme was to consider relatively minor incidents and injuries, as this would be the primary type of incident reported to UPLOADS.

The analysis included 274 de-identified incident reports provided by three Australian LOA providers, and 1014 incident reports from the NZ NID provided by the Mountain Safety Council. The types of contributory factor identified from the analysis of the Australian and NZ incident reports are shown in Figures 6.5 and 6.6, respectively.

6.4.3.4 The Prototype Contributory Factor Classification Scheme

The contributory factors identified from the data sources were combined to create the prototype UPLOADS contributory factor classification scheme, shown in Figure 6.7. The prototype classification scheme consists of three levels of categories. *Level 1* describes the 'LOA system' in terms of ten categories describing the activity context; the key people involved in the activity; and the people and agencies that impact on how the activity is run. *Level 2* breaks the first level categories down into 55 descriptive categories. *Level 3* breaks the second level categories down into between 2 and 19 highly specific contributory factors, giving a total of 325 highly specific categories.

An example of the hierarchical structure and the level of detail captured by the highly specific factors within the classification scheme is presented in Figure 6.8. The three-level structure was devised because incident reports contain varying amounts of detail; therefore, it is not always possible to pin-point a highly specific contributory factor. For example, '*There was a problem with the equipment*' can only be classified as Level 1: Equipment, while a more specific description of the problem, '*Participant only brought thongs for the bushwalk*' can be classified as Equipment: Clothing and PPE: Lack of equipment.

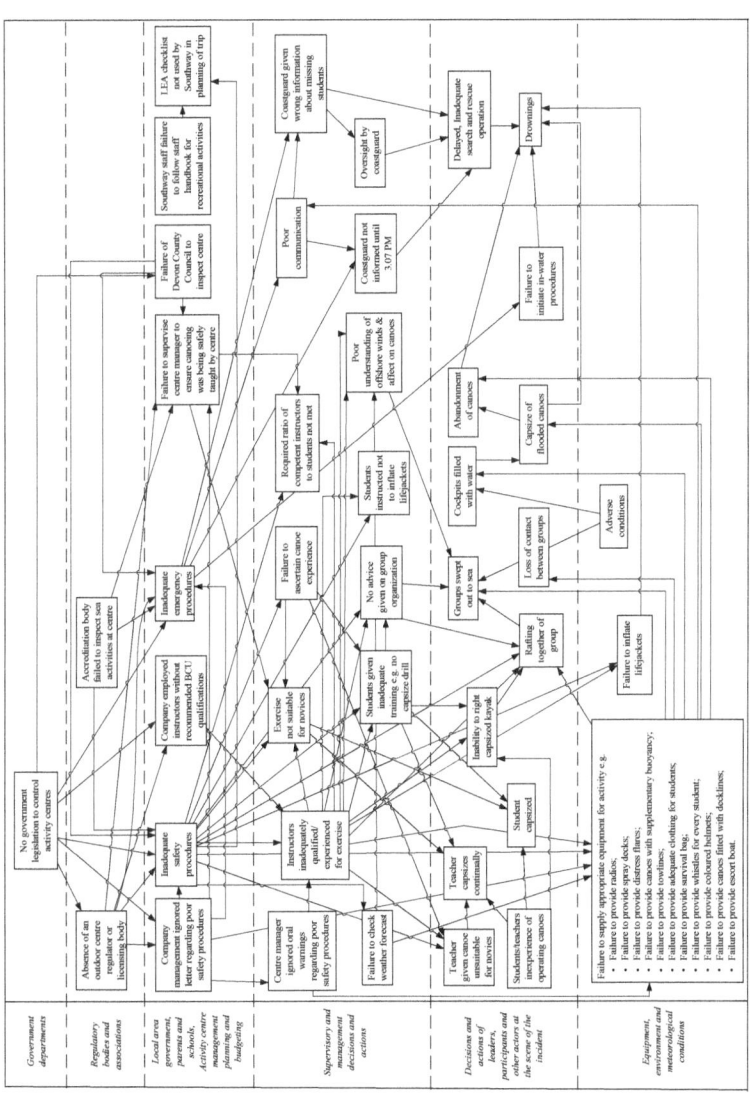

FIGURE 6.3

Contributory factors and relationships identified from the analysis of the Lyme Bay incident. (From Salmon, P. M., Williamson, A., Lenné, M., Mitsopoulos-Rubens, E., & Rudin-Brown, C. M. 2010. *Ergonomics*, 53(8), 927–939. With permission.)

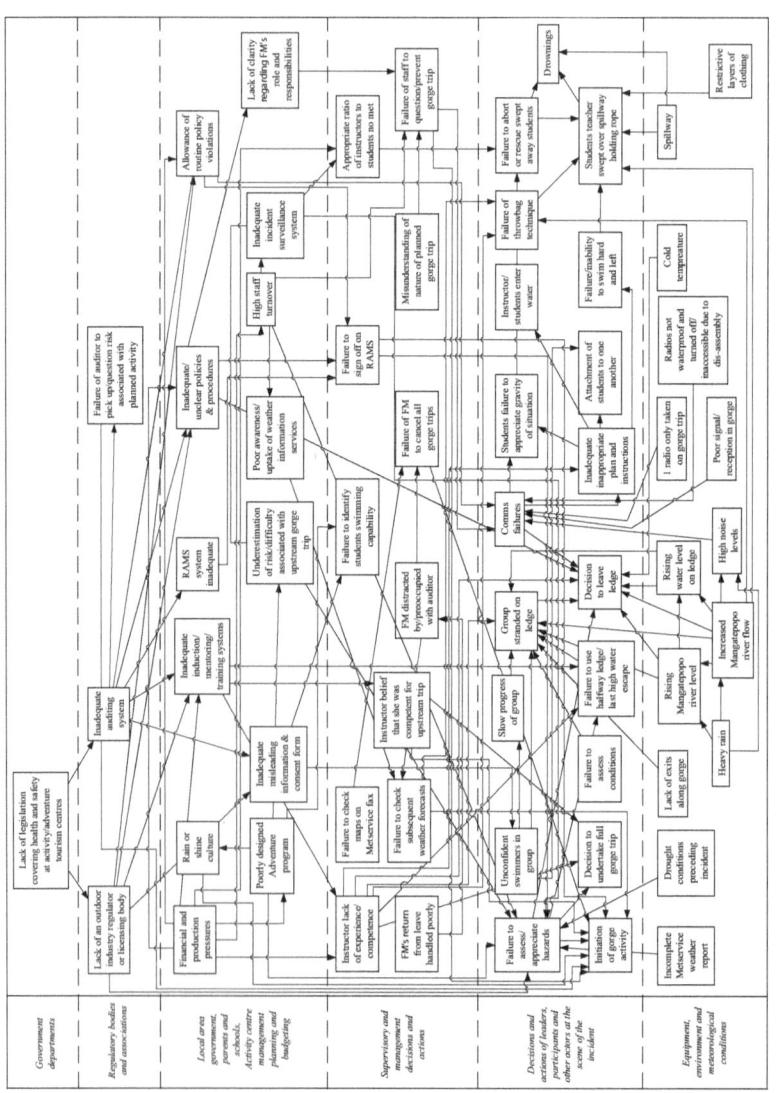

FIGURE 6.4

The contributory factors, and the relationships between them, identified from the analysis of the Mangatepopo Gorge walking incident. (From Salmon, P. M., M. Cornelissen, & M. J. Trotter 2012. *Safety Science* 50 (4):1158–1170. With permission.)

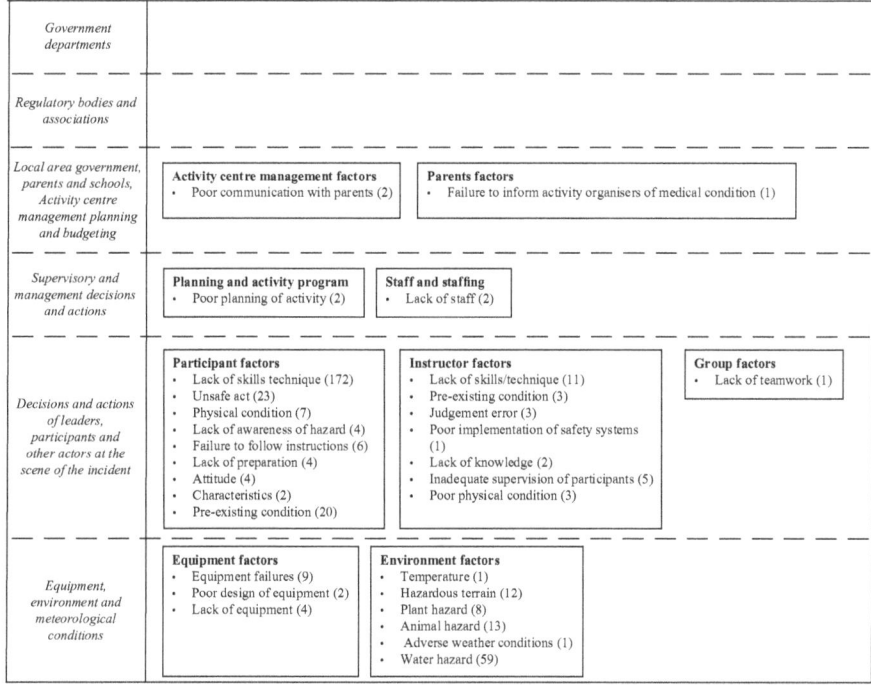

FIGURE 6.5
Types of contributory factors identified from the analysis of incident reports provided by three Australian LOA providers (n = 274 incident reports). Numbers in brackets represents the number of incident reports that identified that type of contributory factor.

Readers should note that the prototype scheme (shown in Figures 6.7 and 6.8) is far from a perfect example of the principles outlined in the Section 6.4.1. The scheme included many 'other' categories and negatively framed codes (e.g. lack of equipment and inadequate equipment). As a result, many of cycles of testing and refinement were required to achieve satisfactory reliability and validity (see Chapter 7).

6.5 Summary and Next Steps

This chapter has provided guidance on adapting Accimap for use within an incident reporting system, using UPLOADS as an example. The adaptation of Accimap for UPLOADS drew on several data sources, including workshops with subject matter experts, the literature, case studies of fatal incidents, and Australian and New Zealand incident reports. This resulted

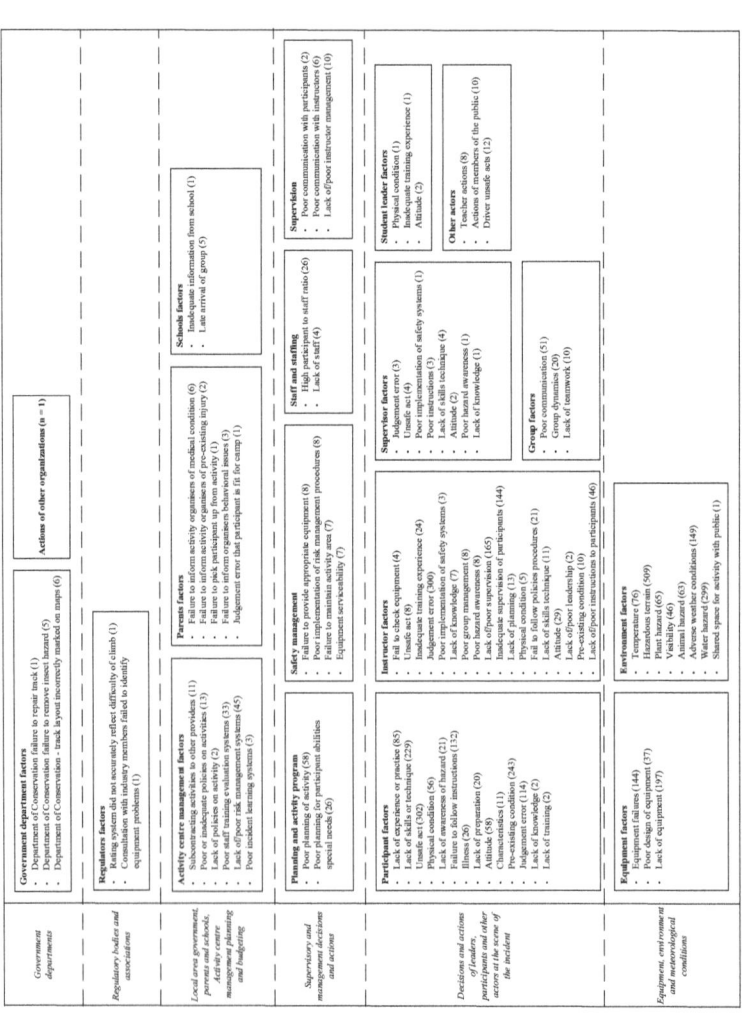

FIGURE 6.6

Types of contributory factors identified from the analysis of incident reports from the NZ NID (n = 1014 incident reports). Numbers in brackets represent the number of incident reports that identified that type of contributory factor. (From Salmon, P. M., M. Cornelissen, & M. J. Trotter 2012. *Safety Science* 50 (4):1158–1170. With permission.)

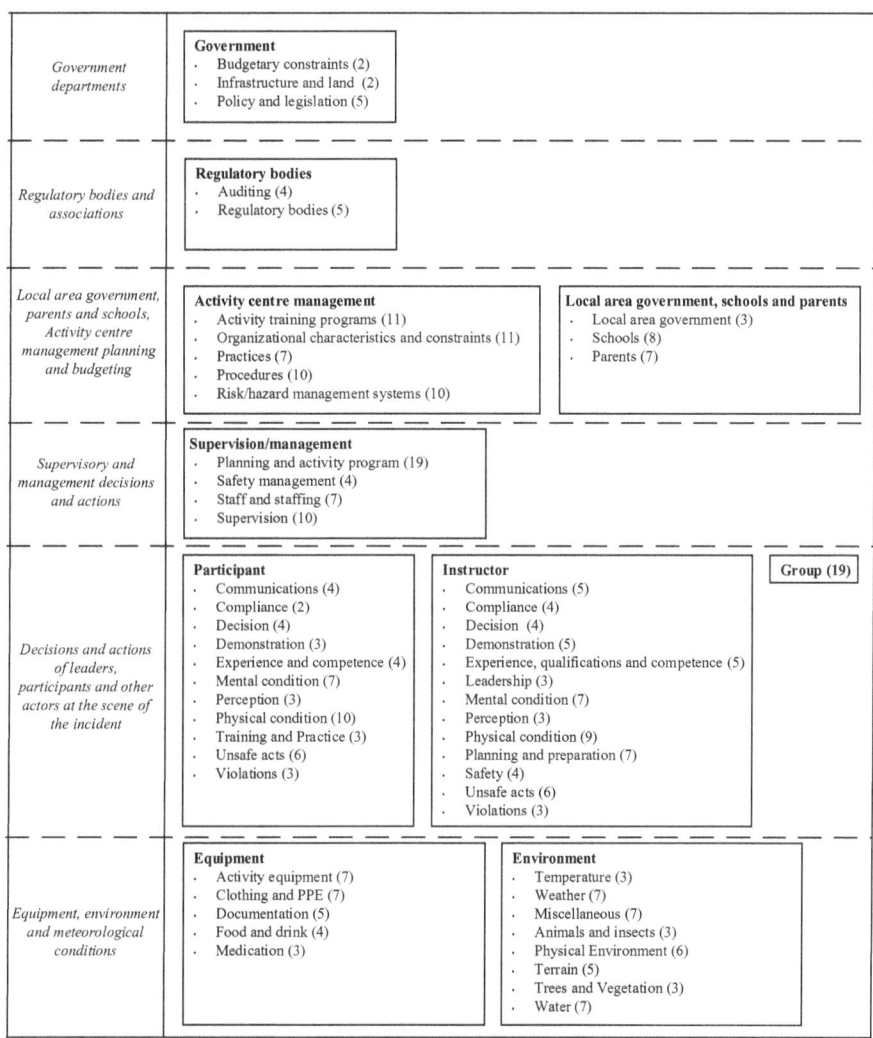

FIGURE 6.7
The UPLOADS accident analysis method with prototype classification scheme. Level 1 codes are shown in bold, with Level 2 codes presented below. The number of corresponding Level 3 codes are indicated in brackets. (From Goode, N., P. M. Salmon, N. Z. Taylor, M. G. Lenné, & C. F. Finch 2017. *Applied Ergonomics* 64:14–26. With permission.)

in an adapted Accimap framework and a highly detailed contributory factor classification scheme that was informed by a wide range of inputs. The process required to test and refine a contributory factor classification scheme to ensure that it is reliable and valid is described in the following chapter.

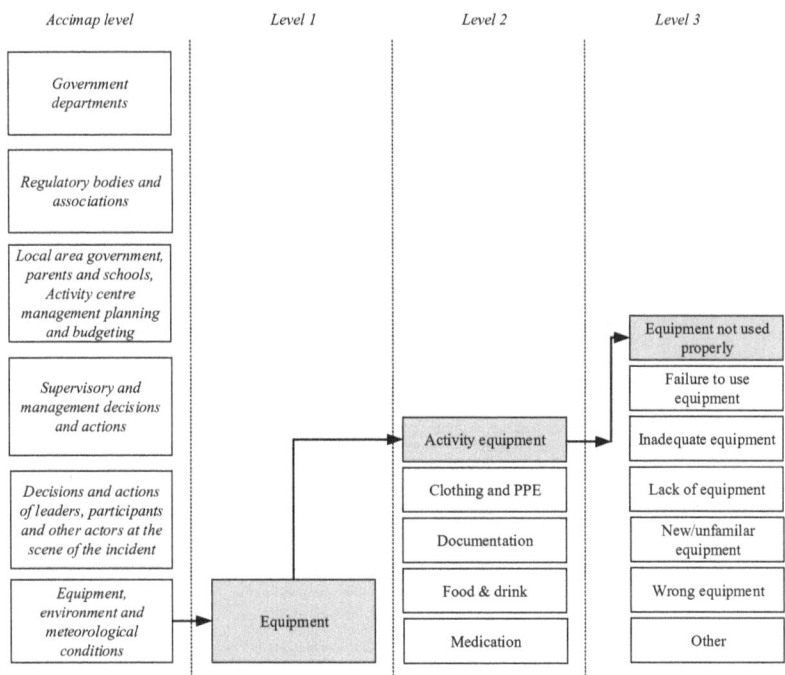

FIGURE 6.8
Example of the hierarchical structure of the prototype classification scheme. (From Goode, N., P. M. Salmon, N. Z. Taylor, M. G. Lenné, & C. F. Finch 2017. *Applied Ergonomics* 64:14–26. With permission.)

Further Reading

Wallace and Ross's (2006a) book, *Beyond Human Error*, provides a comprehensive overview of the theoretical considerations involved in developing safety-related classification schemes.

References

Brookes, A., M. Smith, & B. Corkill. (2009). Report to the Trustees of the Sir Edmund Hillary Outdoor Pursuit Centre of New Zealand: Mangatepopo Gorge Incident, 15th April 2008.
Davenport, C. J. (2010). Mangatepopo coroner's report. Auckland.

Goode, N., P. M. Salmon, N. Z. Taylor, M. G. Lenné, & C. F. Finch. (2017). Developing a contributing factor classification scheme for Rasmussen's AcciMap: Reliability and validity evaluation. *Applied Ergonomics* 64:14–26.

Gordon, R., R. Flin, & K. Mearns. (2005). Designing and evaluating a human factors investigation tool (HFIT) for accident analysis. *Safety Science* 43 (3):147–171. doi: http://dx.doi.org/10.1016/j.ssci.2005.02.002

Jenkins, S., & P. Jenkinson. (1993). Report into the Lyme Bay canoe tragedy. *Devon County Council Report.*

Le Coze, J.-C. (2010). Accident in a French dynamite factory: An example of an organisational investigation. *Safety Science* 48(1):80–90.

O'Connor, P. (2008). HFACS with an additional layer of granularity: Validity and utility in accident analysis. *Aviation Space & Environmental Medicine* 79 (6):599–606. doi: https://doi.org/10.3357/asem.2228.2008

O'Connor, P., J. Walliser, & E. Philips. (2010). Evaluation of a human factors analysis and classification system used by trained raters. *Aviation, Space, and Environmental Medicine* 81 (10):957–960. doi: https://doi.org/10.3357/asem.2843.2010

Olsen, N. S., & S. T. Shorrock. (2010). Evaluation of the HFACS-ADF safety classification system: Inter-coder consensus and intra-coder consistency. *Accident Analysis & Prevention* 42 (2):437–444. doi: http://dx.doi.org/10.1016/j.aap.2009.09.005

Rasmussen, J. (1997). Risk management in a dynamic society: A modelling problem. *Safety Science* 27 (2/3):183–213.

Salmon, P. M., M. Cornelissen, & M. J. Trotter. 2012. Systems-based accident analysis methods: A comparison of Accimap, HFACS, and STAMP. *Safety Science* 50 (4):1158–1170.

Salmon, P. M., N. Goode, M. G. Lenné, C. F. Finch, & E. Cassell. (2014). Injury causation in the great outdoors: A systems analysis of led outdoor activity injury incidents. *Accident Analysis & Prevention* 63:111–120. doi: http://dx.doi.org/10.1016/j .aap.2013.10.019

Salmon, P. M., G. H. Walker, G. Read, N. Goode, & N. A. Stanton. (2016). Fitting methods to paradigms: Are Ergonomics methods fit for systems thinking? *Ergonomics*:1–12.

Salmon, P. M., A. Williamson, M. Lenné, E. Mitsopoulos-Rubens, & C. M. Rudin-Brown. (2010). Systems-based accident analysis in the led outdoor activity domain: Application and evaluation of a risk management framework. *Ergonomics* 53 (8):927–939. doi: 10.1080/00140139.2010.489966

Svedung, I., & J. Rasmussen. (2002). Graphic representation of accident scenarios: Mapping system structure and the causation of accidents. *Safety Science* 40 (5): 397–417.

Wallace, B., & A. J. Ross. (2006a). *Beyond human error: Taxonomies and safety science.* Boca Raton, FL: CRC Press.

Wallace, B., & A. J. Ross. (2006b). Safety and taxonomies. In *Beyond Human Error: Taxonomies and Safety Science*, 31–64. Boca Raton, FL: CRC Press.

Waterson, P., D. P. Jenkins, P. M. Salmon, & P. Underwood. (2016). 'Remixing Rasmussen': The evolution of Accimap within systemic accident analysis. *Applied Ergonomics.*

Wiegmann, D. A., & S. A. Shappell. (2003). *A Human error approach to aviation accident analysis: The human factors analysis and classification system*: Burlington, VT: Ashgate.

Appendix: Literature Review References

American Camp Association. (2011). *Healthy camp study impact report 2006-2010: Promoting health and wellness among youth and staff through a systematic surveillance process in day and resident camps.* Retrieved from http://www.acacamps.org /sites/default/files/images/education/Healthy-Camp-Study-Impact-Report.pdf

Barst, B., Bialeschki, M. D., & Comstock, R. D. (2008). Healthy camps: Initial lessons on illnesses and injuries from a longitudinal study. *Journal of Experiential Education, 30*(3), 267–270.

Boyes, M. A., & O'Hare, D. (2003). Between safety and risk: A model for outdoor adventure decision making. *Journal of Adventure Education & Outdoor Learning, 3*(1), 63–76. doi:10.1080/14729670385200251

Brackenberg, M. (1997). How safe are we? A review of injury and illness in outdoor education programmes. *Journal of Adventure Education and Outdoor Leadership, 14*(1).

Brackenberg, M. (1999). Learning from our mistakes – Before it's too late. *Australian Journal of Outdoor Education, 3*(27–33).

Brookes, A. (2003a). Outdoor education fatalities in Australia 1960–2002. Part 1. Summary of incidents and introduction to fatality analysis. *Australian Journal of Outdoor Education, 7*(2), 20–35.

Brookes, A. (2003b). Outdoor education fatalities in Australia 1960-2002. Part 2. Contributory circumstances: Supervision, first aid, and rescue. *Australian Journal of Outdoor Education, 7*(2), 34–42.

Brookes, A. (2004). Outdoor education fatalities in Australia 1960-2002. Part 3. Environmental circumstances. *Australian Journal of Outdoor Education, 8*(1), 44–56.

Brookes, A. (2007a). Preventing death and serious injury from falling trees and branches. *Australian Journal of Outdoor Education, 11*(2), 50–59.

Brookes, A. (2007b). Research update: Outdoor education fatalities in Australia. *Australian Journal of Outdoor Education, 11*(1), 3–9.

Brookes, A. (2011). Research update 2010: Outdoor education fatalities in Australia. *Australian Journal of Outdoor Education, 15*(1), 35.

Brown, T. J. (1995). Adventure risk management: A practical model. *Australian Journal of Outdoor Education, 1*(2), 16–24.

Capps, K. (2007). *Factors related to the occurrence of incidents in adventure recreation programs.* (Thesis submitted to the Graduate Faculty of North Carolina State University in partial fulfilment of the requirements for a master's degree in science.)

Cline, P. B. (2007). *Learning to interact with uncertainty.* Paper presented at the inaugural Outdoor Education Australia Risk Management Conference, Ballarat, Victoria.

Crosby, S., & Benseman, J. (2003). Vivisecting the Tui: The qualifications framework and outdoor leadership development in New Zealand. *Australian Journal of Outdoor Education, 7*(2), 43–52.

Curtis, R. (1995a). *OA guide to outdoor safety management.* Retrieved from http://www.princeton.edu/~oa/files/safeman.pdf

Curtis, R. (1995b). *OA guide to planning a safe river trip.* Retrieved from http://www.princeton.edu/~oa/paddle/rivplan.shtml. Access date: 15th May 2018.

Dallat, C. (2007). *Communicating risk: An insight into, and analysis of, how some of our colleagues are currently communicating risk to parents.* Paper presented at the Wilderness Risk Management Conference, Banff, Canada.

Dallat, C. (2009). Communicating risk with parents: Exploring the methods and beliefs of outdoor education co-ordinators in Victoria, Australia. *Australian Journal of Outdoor Education, 13*(1), 3(13).

Davidson, G. (2004). Fact or folklore? Exploring 'myths' about outdoor education accidents: Some evidence from New Zealand. *Journal of Adventure Education & Outdoor Learning, 4,* 13–37.

Davidson, G. (2005). *Towards understanding the root causes of outdoor education incidents.* University of Waikato, New Zealand.

Davidson, G. (2006). Fact or folklore? Exploring 'myths' about outdoor education accidents: Some evidence from New Zealand. *New Zealand Journal of Outdoor Education: Ko Tane Mahuta Pupuke, 2*(1), 50–85.

Davidson, G. (2007, 20–23 September 2007). *Towards understanding the root causes of outdoor education incidents.* Paper presented at the 15th National Outdoor Education Conference, Ballarat, Victoria.

Dickson, T. J. (2001). Calculating risks: Fine's mathematical formula 30 years later. *Australian Journal of Outdoor Education, 6*(1), 31(39).

Dickson, T. J., Chapman, J., & Hurrell, M. (2000). Risk in outdoor activities: The perception, the appeal, the reality. *Australian Journal of Outdoor Education, 4*(2), 10–17.

Elliott, T. B., Elliott, B. A., & Bixby, M. R. (2003). Risk Factors Associated With Camp Accidents. *Wilderness & Environmental Medicine, 14*(1), 2–8.

Ewert, A., Shellman, A., & Glenn, L. (2006). *Instructor traps: What they are and how they impact our decision making and judgement.* Paper presented at the Wilderness Risk Management Conference, Killington, VT.

Fiore, D. C. (2003). Injuries associated with whitewater rafting and kayaking. *Wilderness & Environmental Medicine, 14*(4), 255–260.

Forde, L., & Solomonidis, R. (2010). Risky business: Managing off-campus activities. *Teacher: The National Education Magazine,* (April 2010), 54.

Fulbrook, J. (2005). *Outdoor activities, negligence, and the law.* Hampshire: UK: Ashgate.

Galloway, S. (2002). Theoretical cognitive differences in expert and novice outdoor leader decision making: Implications for training and development. *Journal of Adventure Education and Outdoor Learning, 2*(1), 19–28.

Garvey., D. (1998). Risk management: An international perspective. *Journal of Experiential Education, 21*(2), 63–70.

Gray, S., & Moeller, K. (2010). *Behavioral risk management: Preventing critical incidents in the field.* Paper presented at the Wilderness Risk Management Conference, Colorado Springs, CO, USA.

Haddock, C. (1999). *High potential incidents – Determining their significance: Tools for our trade and a tale or two.* Paper presented at the Proceedings of the 1999 Wilderness Risk Managers Conference, Lander, WY, USA: The National Outdoor Leadership School.

Haddock, C. (2004). *Outdoor safety: Risk management for outdoor leaders.* Wellington, NZ: New Zealand Mountain Safety Council.

Haddock, C. (2008). Incident reviews: Aspects of good practice. *Ki Waho – Into the Outdoors, 2.*

Hamonko, M. T., McIntosh, S. E., Schimelpfenig, T., & Leemon, D. (2011). Injuries related to hiking with a pack during National Outdoor Leadership School courses: A risk factor analysis. *Wilderness & Environmental Medicine, 22*(1), 2–6.

Harper, N., & Robinson, D. W. (2005). Outdoor adventure risk management: Curriculum design principles from industry and educational experts. *Journal of Adventure Education and Outdoor Learning, 5*(2), 145–158.

Hogan, R. (2002). The crux of risk management in outdoor programs – Minimising the possibility of death and disabling injury. *Australian Journal of Outdoor Education, 6*(2), 71(76).

Hovelynck, J. (1998). Learning from accident analysis: The dynamics leading up to a rafting accident. *Journal of Experiential Education, 21*(2), 89–95.

Hunter, I. R. (2007). An analysis of white water rafting safety data: Risk management for programme organizers. *Journal of Adventure Education and Outdoor Learning, 7*(1), 21–35.

Isaac, A. (1997). The cave creek incident: A REASONed explanation. *Australasian Journal of Disaster and Trauma Studies, 3.*

Le Coze, J.-C. (2010). Accident in a French dynamite factory: An example of an organisational investigation. *Safety Science, 48*(1), 80–90.

Martin, B., Cashel, C., Wagstaff, M., & Breunig, M. (2006). *Outdoor leadership: Theory and practice.* USA: Human Kinetics.

Meyer, D. (1979). The management of risk. *Journal of Experiential Education, 2*(2), 9–14.

Powell, C. (2007). The perception of risk and risk taking behaviour: Implications for incident prevention strategies. *Wilderness and Environmental Medicine, 18,* 10–15.

Salmon, P. M., Cornelissen, M., & Trotter, M. J. (2012). Systems-based accident analysis methods: A comparison of Accimap, HFACS, and STAMP. *Safety Science, 50*(4), 1158–1170.

Salmon, P. M., Williamson, A., Lenné, M., Mitsopoulos-Rubens, E., & Rudin-Brown, C. M. (2010). Systems-based accident analysis in the led outdoor activity domain: Application and evaluation of a risk management framework. *Ergonomics, 53*(8), 927–939. doi:10.1080/00140139.2010.489966

Watters, R. (1996). *Whitewater river accident analysis.* Paper presented at the Proceedings of 1995 International Conference on Outdoor Recreation and Education.

Wilks, J., & Davis, R. J. (2000). Risk management for scuba diving operators on Australia's Great Barrier Reef. *Tourism Management, 21,* 591–599.

7

Evaluating Reliability and Validity

Practitioner Summary

Using an unreliable or invalid contributory factor classification scheme in an incident reporting system would mean that the findings would provide a misleading picture of incident causation. This chapter provides step-by-guidance on evaluating the reliability and validity of a contributory factor classification scheme, and then refining the scheme to achieve satisfactory levels of reliability and validity. The application of this process is illustrated by describing an evaluation of the UPLOADS contributory factor classification scheme. This provides a detailed example of how to conduct, report, and interpret the findings from such an evaluation. The chapter concludes with a discussion of how the evaluation findings informed the subsequent design of the UPLOADS software tool and training materials, which is relevant to the design of incident reporting systems in any domain.

Practical Challenges

Developing a classification scheme that is reliable, valid, and sufficiently detailed for detecting problems is a difficult task. The more categories of contributory factors that are included in the scheme, the more opportunities there are for analysts to misinterpret them. These difficulties can only be resolved through repeated cycles of testing and refinement.

7.1 Introduction

Within an incident reporting system, a contributory factor classification scheme has two primary purposes. First, the scheme should ensure that analyses of incident reports adequately reflect the underpinning accident causation model by highlighting the types of contributory factors that may have been involved in an incident. Second, the scheme should standardize the analysis of incident reports, so that the same types of contributory factors are classified in the same way across multiple incidents, and by different analysts, to support the consistent aggregation of data. To determine whether a classification scheme is likely to meet these requirements, it is necessary to evaluate its reliability and validity when used to analyze incident reports. Evaluation findings can also provide vital information for refining the scheme so that, ultimately, it is reliable and valid.

This chapter begins by defining reliability and validity, and then presents a step-by-step guide to evaluating and refining a contributory factor classification scheme. The sequence of assessments undertaken to test and refine the UPLOADS classification scheme is then described. The last assessment in the sequence is presented as an example of testing the reliability and validity of a classification scheme of this type. The chapter concludes with a discussion of the findings, and the constraints they placed on the design of the UPLOADS software tool.

7.2 What Is Reliability and Validity?

There are many forms of reliability and validity that are relevant to the development of contributory factor classification schemes (Annett, 2002; Stanton, 2016; Stanton & Young, 1999; Wallace & Ross, 2006b). In the early stages of development, construct and content validity are important, as they indicate that a scheme adequately reflects the underpinning accident causation theory. Once this is established, satisfactory levels of criterion-referenced validity, as well as intra-analyst and inter-analyst reliability need to be achieved prior to implementing the scheme in an incident reporting system.

CRITERION-REFERENCED VALIDITY

Applying the scheme produces the same outputs as a specified criterion.

Criterion-referenced validity refers to whether applying the scheme results in the same outputs as a specified criterion. For example, a human error identification technique is said to be valid if the predicted errors are the same as the actual errors that are observed for a particular task (Stanton & Stevenage, 1998). To evaluate the validity of a contributory factor classification scheme, the categories selected by an individual analyst are usually compared to a 'gold standard', such as an analysis conducted by a highly experienced analyst, an expert panel or the creators of the scheme. A high level of criterion-referenced validity is achieved when the categories selected by analysts from the scheme are the same as those generated by the gold standard analysis. This provides one form of evidence that the scheme can be used as intended (Stanton, 2016; Stanton & Young, 1999, 2003).

Intra-analyst reliability refers to whether an individual analyst chooses the same categories to describe the same contributory factors on repeated occasions (Stanton & Young, 2003). It is often referred to as test-retest reliability. A high level of intra-analyst

INTRA-ANALYST RELIABILITY

The same analyst chooses the same categories for the same contributory factors on repeated occasions.

reliability indicates that an analyst consistently selects the same categories to describe the same contributory factors on repeated occasions. This is a good indication that the classification scheme is simple to understand and use, that categories do not overlap, and that the training material explaining the scheme is clear and easy to understand. A high level of intra-analyst reliability ensures that the use of the scheme and its categories does not vary over time for the same set of contributory factors. This is extremely important for incident reporting systems because incident data is typically collected and analyzed over months and years. It is also important because the same contributory factors will often appear across multiple incidents, and analysts need to categorize them consistently.

Inter-analyst reliability refers to the consistency among two or more different analysts (Stanton & Young, 2003). A high level of inter-analyst reliability indicates that different analysts consistently select the same categories to describe the same con-

INTER-ANALYST RELIABILITY

Different analysts choose the same categories for the same contributory factors.

tributory factors for the same given incident. This indicates that the classification scheme is understood and applied in the same way by different people. It also provides evidence that the contributory factor classification scheme is logically organized and parsimonious, with few overlapping categories (Ross, Wallace, & Davies, 2004), and that any guidance or training material is clear and easy to follow. A high level of inter-analyst reliability is particularly important when the classification scheme is included as a checklist on an incident report (i.e. many different people must use it when they report an incident), or when many people within a safety department contribute to analyzing incidents.

7.3 Step-by-Step Guide to Evaluating Reliability and Validity

The aim of the process described in this section is to evaluate the intra- and inter-analyst reliability and the criterion-referenced validity of a contributory factor classification scheme, and then (if required) refine the scheme based on the results.

A flowchart describing the phases of the process is shown in Figure 7.1. The process uses a test-retest design to evaluate intra- and inter-analyst reliability in the same assessment. In this type of assessment, all participants complete the same pre-determined coding tasks on two separate occasions (Time 1 and Time 1). The time between coding tasks needs to be long enough (i.e. at least 2 to 4 weeks) so that participants are not able to complete the tasks from memory on the second occasion.

The following sections describe the activities that need to be completed during each phase of the process, and the factors that need to be considered to ensure that the assessment is robust.

7.3.1 Step 1: Identify and Recruit Participants for the Assessment

To assess intra- and inter-analyst reliability, it is generally recommended that six to eight 'novice' participants complete the assessment (Wallace & Ross, 2006a). In our own studies, we have found that this number of participants is sufficient for detecting poorly defined and overlapping categories. There tends to be diminishing returns in including more participants.

Novice participants can be drawn from a range of backgrounds – it is not necessary to include end users in every assessment. Researchers, human factors specialists, and students can participate in the early stages of development to identify obvious problems with the classification scheme. Later stages of development, and especially prior to implementation, should involve end users to ensure that the scheme is reliable and valid when used by them.

To provide a reference point for assessing criterion referenced validity, the coding tasks also need to be completed by at least two 'experts'. The experts should either be highly experienced analysts in using the scheme, or the creators of the scheme.

7.3.2 Step 2: Design the Training

Training can be provided in several ways, including: face to face; as written material including worked examples; or in an online tutorial.

It is essential that the training provided for the assessment, is the same as the training that will ultimately be provided when the scheme is used as part of the incident reporting system. Any training where participants are allowed to practice using the classification scheme on a relevant incident, discuss the selection of categories, ask questions, or receive feedback will obviously result in better reliability and validity than without this assistance (Olsen, 2013). If the training provided during the assessment is much more extensive than what will be provided as part of the incident reporting system, then the assessment will produce inflated results.

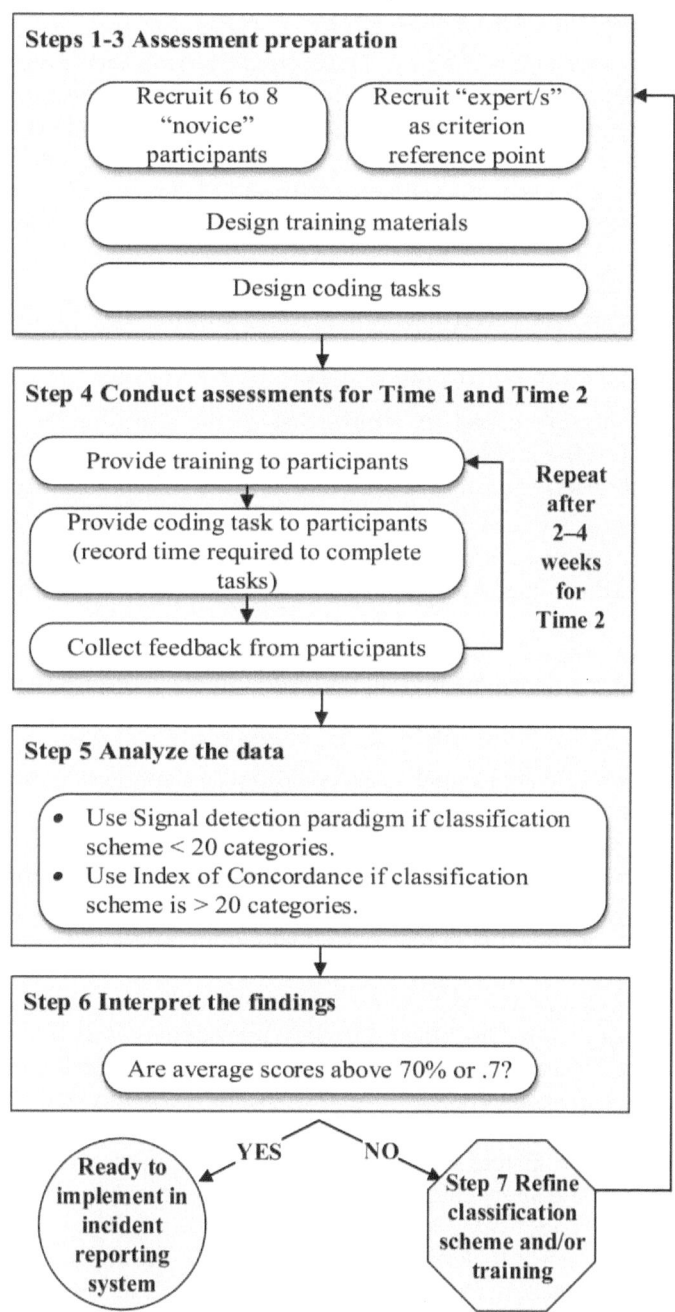

FIGURE 7.1
Flowchart illustrating the phases of an assessment designed to evaluate intra- and inter-analyst reliability and criterion referenced validity. The cycle is repeated until satisfactory reliability and validity are achieved.

7.3.3 Step 3: Design the Coding Tasks

After analysts receive training on the scheme, they are then given a series of coding tasks to perform independently where they use the classification scheme to categorize the contributory factors from incident reports.

A coding task typically starts with the presentation of an incident narrative, which includes a qualitative description of the contributory factors involved in the incident. There are then three different ways to present the coding task:

1. Participants are given a checklist representing the classification scheme, and instructed to select all the categories that are relevant to the incident report (O'Connor, 2008).
2. Participants are asked to write a list of the contributory factors involved in the incident, and then to select a category, or categories, that best describe each factor (Gordon, Flin, & Mearns, 2005).
3. Participants are presented with a list of pre-identified contributory factors from the incident report, along with a space beside each for selecting category, or categories, that best describe each factor (Olsen, 2011).

An example incident narrative is presented in Appendix B, with examples of corresponding coding tasks presented using these three different approaches.

The type of coding task that is given to participants will influence the conclusions that can be drawn from the assessment. The first two types of coding tasks described above mirror how classification schemes are often deployed within incident reporting systems; however, they are not optimal for diagnosing specific problems with the classification scheme. The problem with using a checklist is that participants may select the same categories but actually identify different contributory factors (Ross et al., 2004). Similarly, if participants must identify contributory factors from the report *and* select categories to describe them, low inter-analyst reliability may be attributable to disagreements over the relevancy of the contributory factor, the selection of the category from the classification scheme, or both (Olsen, 2013; Ross et al., 2004; Wallace & Ross, 2006a). The third option – presenting a list of pre-defined factors – allows for an examination of whether analysts consistently use the same categories to describe the same contributory factors. Problematic codes can then be easily identified.

An additional factor to consider when designing the coding tasks is the selection of appropriate incident reports or narratives. Ideally, the incident reports included in the assessment would utilize all the categories within the classification scheme; however, this is often not possible due to the time it would then take to conduct the assessment. For a classification scheme underpinned by Accimap, the narratives should at least utilize categories from all levels of the framework. For example, in the case of the prototype UPLOADS classification scheme, this would include contributory factors that could be classified using all six levels in Figure 7.2. It would be a poor test of this scheme if the incident reports only referred to contributory factors involving activity leaders, equipment, and the activity environment.

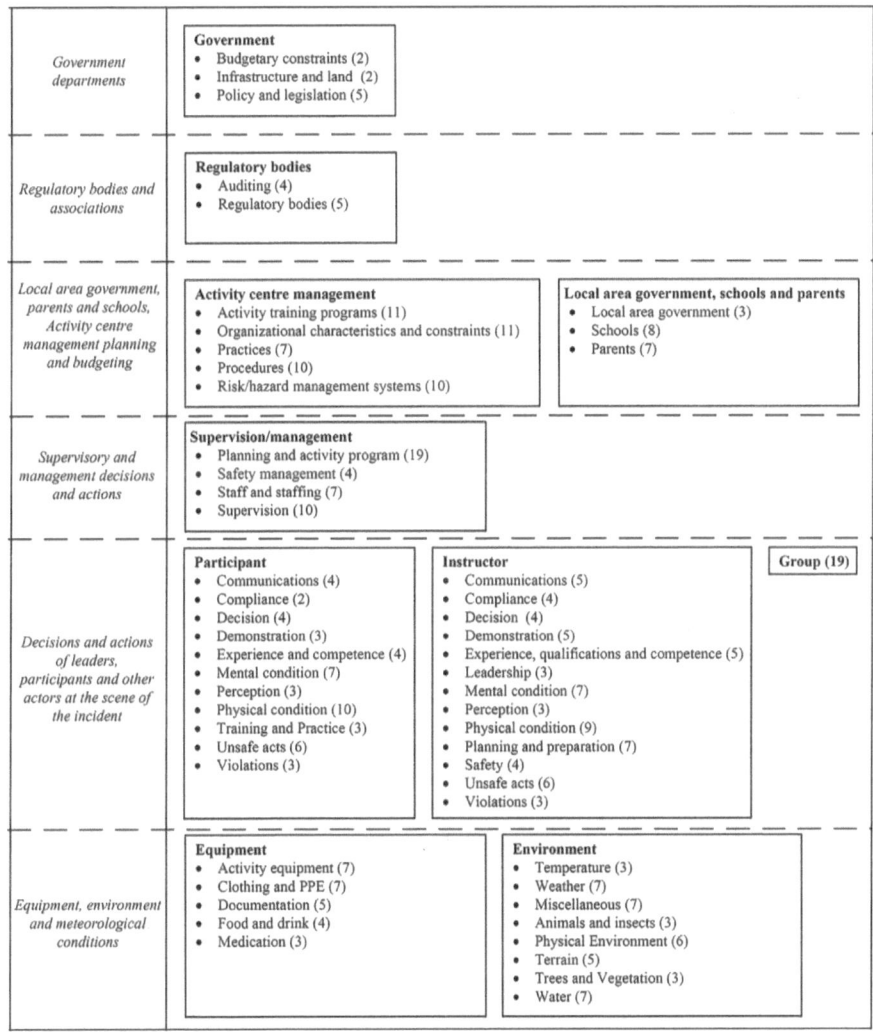

FIGURE 7.2
The UPLOADS prototype classification scheme. Level 1 categories are shown in bold, with Level 2 categories presented below. The number of corresponding Level 3 categories are indicated in brackets. (From Goode, N., Salmon, P. M., Taylor, N. Z., Lenné, M. G., & Finch, C. F. 2017b. *Applied Ergonomics, 64,* 14–26. With permission.)

7.3.4 Step 4: Conduct the Assessments

The assessments can be conducted via email, post, or face to face, depending on the design of the training.

As shown in Figure 7.1, participants should be provided with the training, and then given the coding tasks to complete (Time 1). The time taken to complete the coding tasks should recorded, and the responses collected as soon as the tasks are completed. It is also useful to collect feedback from participants on their perceptions of the scheme. This procedure should be repeated in exactly the same way at Time 2. During the intervening time, participants should not receive any feedback or clarification on the coding tasks.

7.3.5 Step 5: Data Analysis

Once the assessment has been conducted, the patterns of categories selected by each participant need to be analyzed to determine whether satisfactory reliability and validity has been achieved. The most common metrics used to calculate reliability and validity are:

- The percentage agreement for each category in the classification scheme;
- The Index of Concordance (also referred to as a raw agreement); and
- The signal detection paradigm.

Each metric has strengths and weaknesses for evaluating contributory factor classification schemes, as shown in Table 7.1. Based on these strengths and weaknesses, the signal detection paradigm is the best approach for classification schemes with a low number of categories (i.e. under 20), and the Index of Concordance is a reasonable approach for schemes with a high number of categories.

To assess intra-analyst reliability, the categories selected at Time 1 by each participant are compared to the categories they selected at Time 2.

To assess inter-analyst reliability, the categories selected by each participant at Time 1 are compared to all other participants for Time 1, and again for Time 2.

To assess criterion-referenced validity, the categories selected by each participant at Time 1 are compared with those selected by the experts for Time 1, and again for Time 2.

7.3.6 Step 6: Interpreting the Findings

There is no universally accepted criterion for satisfactory reliability and validity for any of the metrics presented in Table 7.1, although 70% for the Index of Concordance (Wallace & Ross, 2006b) and .7 for the Sensitivity

TABLE 7.1

Summary of Metrics for Analyzing the Reliability and Validity of Contributory Factor Classification Schemes

Metric	Calculation	Strengths	Weaknesses
Percentage agreement for each code in the classification scheme	For each coding task, calculate for each category in the classification scheme: $$\frac{\text{No. participants who selected the category}}{\text{Total number of participants}} \times 100$$ This produces percentage agreement for each category in the scheme (O'Connor, 2008).	Easiest way to identify categories with the least agreement.	Produces an inflated result as it ignores the total number of categories selected by each participant.
Index of Concordance (percentage agreement)	For each pair of participants, score 'agreement' or 'disagreement' for each pair of categories assigned to each contributing factor in the coding task. Calculate the Index of Concordance for each pair of participants: $$\frac{\text{Agreements}}{\text{Agreements} + \text{Disagreements}} \times 100$$ Average the Index across all participant pairs to produce a percentage for the coding task (Goode et al., 2017a).	Accounts for level of agreement and disagreement. Penalizes schemes with large numbers of overlapp-ing categories.	Does not account for 'chance agreement'.
Signal detection paradigm	For each pair of participants (e.g. P1 and P2), score a 'hit', 'miss', false alarm', or 'correct rejection' for each category in the classification scheme for a coding task. Hit = category is selected by both P1 and P2; Miss = category is selected by P1 but not P2; False alarm (FA) = category is selected by P2 but not P1; and Correct rejection (CR) = category is not selected by P1 or P2. Calculate the sensitivity index (Stanton & Stevenage, 1998) for each the pair of participants: $$\frac{\left(\dfrac{\text{Hits}}{\text{Hits} + \text{Misses}}\right) + 1 - \dfrac{\text{FAs}}{\text{FAs} + \text{CRs}}}{2}$$ Average this score across participants.	Sensitivity index accounts for trade-offs between hits, misses, false alarms, and correct rejections.	Sensitivity is artificially inflated by large numbers of correct rejections in schemes with a large number of categories.

Index (Stanton & Stevenage, 1998) are generally thought to be a minimum standard. It is also worth noting that all the metrics presented in Table 7.1 are highly influenced by the sample size – it is much easier to obtain a score above 70% or .7 in a study involving two participants, as opposed to a study involving 4 or 10 participants.

7.3.7 Step 7: Refine the Classification Scheme (If Required)

If the scores from the assessment are below the 70% or .7 threshold, then this indicates that the classification scheme needs to refined and re-tested until acceptable performance is achieved.

There are several reasons that low reliability and validity scores might be achieved, including:

1. A poorly executed scheme (e.g. irrelevant categories, too many categories, overlapping or ambiguously named categories, and complex categories that are hard to understand);

2. Inadequate training, including poorly written or presented guidance and training materials;

3. Inadequately detailed narratives that are hard to interpret;

4. Participant fatigue due to too many coding tasks;

5. The presence of a rogue participant; and

6. Disparate backgrounds of participants (Finch et al., 2012; Neuendorf, 2016).

Typically, a close examination of the data (e.g. examining scores for each coding task, scores for each category across coding tasks, and the performance of individual participants) and feedback from participants will help identify the source of poor performance, and suggest the next steps required.

If the source of the poor performance appears to be the scheme itself, then revisions to the scheme and its associated training material are required. A useful first step in refining a scheme is checking whether participants have selected many categories to describe the same contributory factor. This indicates that there are overlapping categories. This problem can be addressed by changing the names of categories so they are more precise; expanding the definitions given for categories; or eliminating categories that appear to refer to overlapping concepts.

Eliminating categories can increase reliability, but it may reduce the usefulness of the scheme.

Although eliminating categories that show poor reliability is often tempting, this needs to be carefully balanced against: (a) ensuring that the scheme comprehensively describes the contributory factors involved in incidents; and (b) preserving the usefulness of the scheme. It is recommended that the labels for the categories and definitions are refined prior to eliminating categories.

7.4 Evaluating the UPLOADS Classification Scheme

The previous section presented a step-by-step guide to evaluating and refining a contributory factor classification scheme. In this section, we demonstrate the application of this process through the development of the UPLOADS classification scheme. Testing and refining the UPLOADS classification scheme and associated training material involved seven assessment cycles, as shown in Figure 7.3.

The prototype UPLOADS contributory factor classification scheme (see Figure 7.2, see p. 121) was initially refined through testing by its creators, other human factors researchers, and then LOA practitioners (Cycles 1 to 4 in Figure 7.3). The prototype scheme had a hierarchical structure with three levels of categories, with a total of 325 highly specific categories at Level 3. The Level 3 categories reflected the level of detail that was identified in pre-existing incident reports.

The first assessment of the prototype scheme with LOA practitioners (Cycle 4), however, showed that the inter-analyst reliability of the prototype scheme was very low (Goode, Salmon, Lenné, & Finch, 2014). Only Level 1 of the scheme showed acceptable levels of inter-analyst reliability, and participants often selected many Level 3 categories to describe the same factor. Feedback from participants indicated that it also required over an hour to analyze each incident report due to the large number of Level 3 categories. It was concluded that the Level 2 categories might represent a reasonable trade-off between detail and reliability, although revisions were still required to ensure the Level 2 categories were mutually exclusive.

The classification scheme was revised in light of the findings from Cycle 1 through 4, and then reduced to two levels of categories (see Figure 7.4). Level 1 describes the 'LOA system' in terms of 14 categories describing the actors, artefacts, and activity context. Level 2 describes specific contributing factors relating to each of these components, with a total of 107 categories. The revised scheme then underwent further cycles of testing and refining the associated definitions and training materials (Cycle 5 to 7). The final version of the scheme and training materials are significantly different from the prototype version – this illustrates the value and importance of the seven cycles of testing.

The following section describes the final assessment as an example of how to conduct, report, and interpret the findings from such an evaluation.

7.4.1 UPLOADS Reliability and Validity Evaluation

The aim of the assessment was to determine whether LOA practitioners could appropriately use the revised classification scheme shown in Figure 7.4. The assessment involved evaluating the intra- and inter-analyst reliability and criterion-referenced validity for each level of the classification scheme.

7.4.1.1 *Procedure and Materials Used in the Assessment*

Eleven LOA practitioners who held a role in their organization with direct responsibility for safety management (e.g. the risk manager,

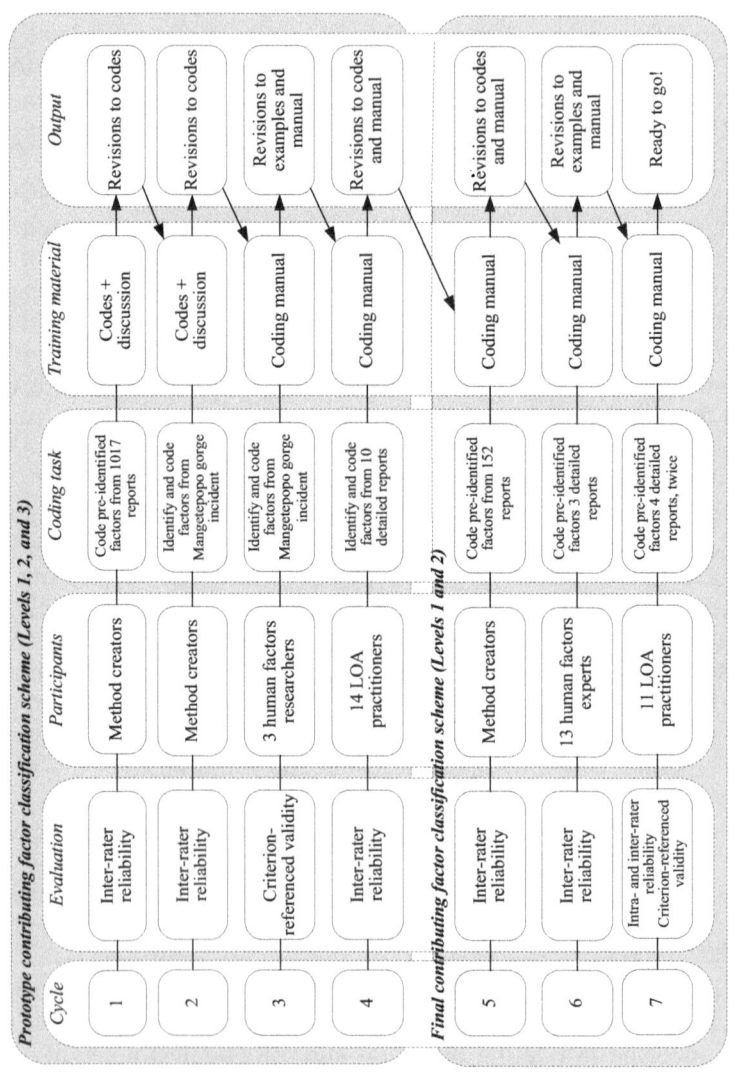

FIGURE 7.3

Assessment cycles required to test and revise the UPLOADS classification scheme and associated training material until satisfactory levels of reliability and validity were achieved.

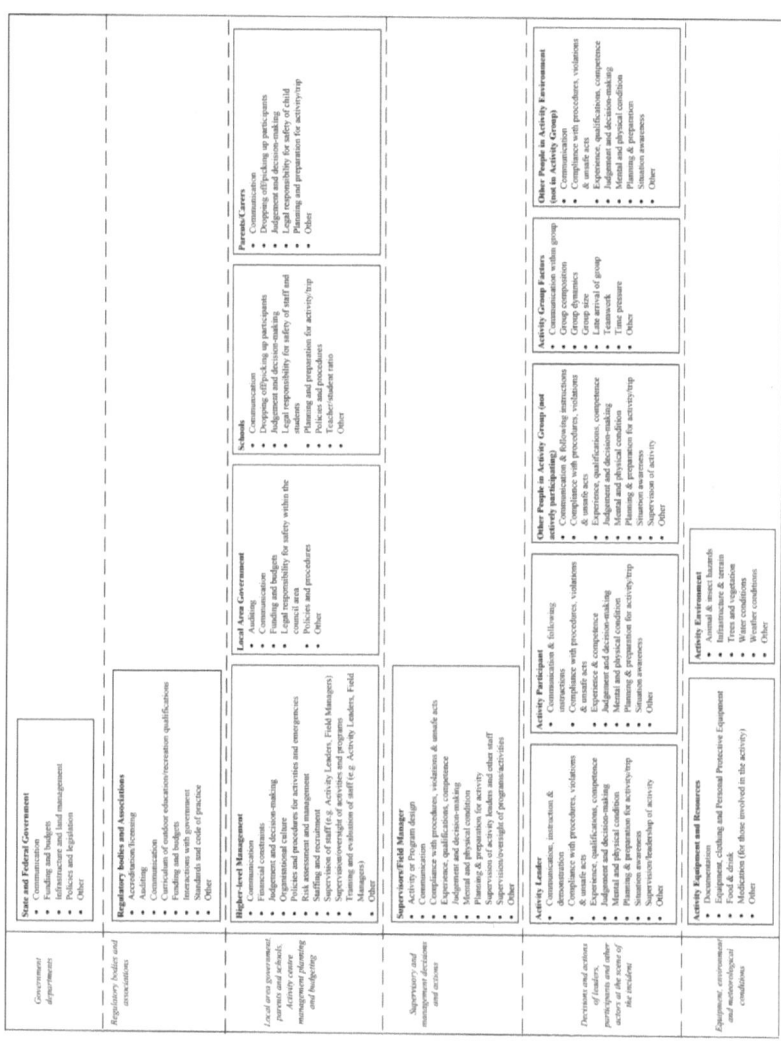

FIGURE 7.4

The UPLOADS classification scheme. Level 1 codes are shown in bold. Level 2 codes are presented below each corresponding system level code. (From Goode, N., Salmon, P. M., Taylor, N. Z., Lenné, M. G., & Finch, C. F. 2017b. *Applied Ergonomics, 64*, 14–26. With permission.)

program manager, senior teacher, and director of outdoor education) participated in the assessment. On average, participants had 17 years of experience
in the LOA sector.

One researcher, who was not involved in the development of the method,
provided instructions, training manuals, and coding booklets to the analysts via email. For the first coding (Time 1), the training manual and coding
booklets were emailed to participants, with a request for them to email the
completed booklets back to the same email address within a fortnight. For
the second coding (Time 2), the same training material and coding booklet
was emailed a month after they had returned their first coding booklet, with
a request to return the completed analyses within a fortnight. Participants
were instructed not to refer to their earlier analyses, and to read through the
training manual again. Participants did not receive any feedback or clarifications regarding the training manual or classification scheme during the
assessment.

The training manual presented a brief overview of the underpinning theoretical framework (Rasmussen, 1997), and described the hierarchical structure of the classification scheme (see Appendix A for a full description) with
examples taken from the analysis of over 1,000 incident reports which we
had used to develop the classification scheme (see Chapter 6). These examples were therefore reflective of our understanding of the scheme. Finally,
several clarifications were presented in the manual based on issues identified in previous assessments.

The coding booklet consisted of four detailed incident reports with a list
of pre-identified contributory factors along with a space beside each for participants to enter the appropriate categories from the classification scheme.
An example of one of the incident narratives with pre-identified contributory
factors is presented in Appendix B. The coding booklet included the instructions not to go beyond the details presented in the reports, and to select the
categories that best described each contributory factor.

Two of the method creators independently completed the coding booklets.
There was approximately 95% agreement regarding the classification of the
contributory factors across the four incidents. This information was used
as a reference point to calculate criterion-referenced validity scores for each
participant.

7.4.1.2 Data Analysis

The Index of Concordance was used to calculate scores for intra- and
inter-analyst reliability and validity for Level 1 and 2 of the classification
scheme.

For each pair of participants coding booklets (e.g. Participant 1 at Time 1 vs
Participant 1 at Time 2, etc.), this involved scoring the agreement or disagreement for each pair of categories assigned to each contributory factor. For each
pair of participant coding booklets, the total number of agreements was then

divided by the total number of agreements and disagreements (agreements/ (agreements + disagreements). The mean was then calculated across all pairs of participant coding booklets. A criterion of 70% agreement was used to evaluate whether the reliability and validity of the classification scheme was satisfactory.

7.4.1.3 Results from the Assessment

The intra-analyst reliability scores (shown in Table 7.2) show that most of the participants used the classification scheme in the same way both times they completed the coding tasks. On average, the intra-analyst reliability scores were above the satisfactory threshold when using the Level 1 and 2 categories. All participants, apart from Participant 4, scored above the acceptable threshold when using the Level 1 categories. Seven out of 11 participants scored above the acceptable threshold when using the Level 2 categories.

The inter-analyst reliability scores (shown in Table 7.3) show that the consistency among the analysts was less than satisfactory. For Level 1, the scores were slightly below the satisfactory threshold for Time 1 (68.8%), and slightly above for Time 2 (73.9%). For Level 2, the scores were below the acceptable threshold for both coding attempts, 58.5% and 64.1% for Times 1 and 2, respectively. This suggested that the training and examples for the classification scheme needed to be improved, especially for the Level 2 categories.

Finally, the validity scores (shown in Table 7.4), which compare participants classifications against the method creators classifications, show that

TABLE 7.2

Summary of Intra-Analyst Reliability Scores

Participant	Level 1 (%)	Level 2 (%)
1	93.4	80.2
2	89.4	80.6
3	94.6	92.7
4	69.3	56.9
5	79.0	67.9
6	74.4	72.2
7	79.7	61.7
8	88.4	90.9
9	81.8	73.4
10	83.2	71.9
11	86.3	65.4
Overall	83.6	74.0

Source: Adapted from Goode, N., Salmon, P. M., Taylor, N. Z., Lenné, M. G., & Finch, C. F. 2017b. *Applied Ergonomics, 64,* 14–26. With permission.

TABLE 7.3

Summary of Inter-Analyst Reliability Scores

| | Level 1 | | | | Level 2 | | | |
| | Time 1 | | Time 2 | | Time 1 | | Time 2 | |
Incident	M (%)	SD	M (%)	SD	M (%)	SD	M (%)	SD
1	73.1	9.2	75.3	11.1	57.2	11.5	64.2	11.7
2	66.0	11.9	78.2	9.5	58.2	12.8	64.1	10.2
3	73.8	11.6	71.6	15.3	56.8	15.2	61.9	13.5
4	62.2	9.1	70.3	8.8	61.7	10.6	66.4	10.5
Overall	68.8	10.5	73.9	11.2	58.5	12.5	64.1	11.5

Source: From Goode, N., Salmon, P. M., Taylor, N. Z., Lenné, M. G., & Finch, C. F. 2017b. *Applied Ergonomics, 64,* 14–26. With permission.

TABLE 7.4

Summary of Criterion-Referenced Validity Scores

| | Level 1 | | | | Level 2 | | | |
| | Time 1 | | Time 2 | | Time 1 | | Time 2 | |
Incident	M (%)	SD	M (%)	SD	M (%)	SD	M (%)	SD
1	79.8	10.0	79.4	9.2	65.3	10.3	70.9	12.9
2	73.7	8.4	79.9	9.0	68.4	9.8	71.5	8.8
3	73.1	12.1	70.2	10.5	63.5	18.2	65.2	14.2
4	69.1	10.5	71.7	7.7	73.3	9.1	75.8	11.5
Overall	73.9	10.3	75.3	9.1	67.6	11.9	70.8	11.9

Source: Adapted from Goode, N., Salmon, P. M., Taylor, N. Z., Lenné, M. G., & Finch, C. F. 2017b. *Applied Ergonomics, 64,* 14–26. With permission.

participants used Level 1 of the classification scheme in a similar way to the method creators, and their use of Level 2 of the scheme became more similar to that of the methods creators with repeated use. On average, the validity scores for Level 1 were above the satisfactory threshold for Time 1 (73.9%) and Time 2 (75.3%). On average, validity scores for Level 2 were slightly below the satisfactory threshold for Time 1 (67.6%) and slightly above for Time 2 (70.8%).

7.4.1.4 Key Insights from the Assessment

The most detailed level of the classification scheme (Level 2) did not achieve satisfactory levels of inter-analyst reliability or validity when used by LOA practitioners. One option available was to reduce the classification scheme to only Level 1, which only contained 14 broad categories, however, it was felt that this would greatly reduce the overall utility of the scheme for incident prevention. For example, in the case of multiple incidents involving issues associated with activity or program design, this contributory factor would be classified simply as 'Supervisors/Field Managers'. This would not provide sufficient information for understanding the underlying issues identified in the incident reports.

To address the identified issues, the following measures were incorporated into the design of the UPLOADS software tool and training materials. The primary decision was to limit the number of people who would be involved in the analysis of incident reports within a single organization. The incident report form was designed to collect a detailed qualitative description of the incident, and any potential contributory factors. The supporting software tool would then be used by a safety manager to subsequently identify and classify the contributory factors from the reports, and they would receive training on the application of the scheme. It was considered that a potential benefit of this strategy was that the safety manager could use their organizational experience to add any additional information to the analysis.

A second measure was to incorporate the examples from the written training material into the software tool for analyzing reports. For example, if 'Documentation' was selected in the software tool, then the text 'e.g. lack of/ incorrect maps, participant lists, participant details, consent forms' would be displayed in the user interface. It was intended that the person analyzing the report could use this information to check whether the category selected was appropriate.

A third measure was the development of several worked examples for inclusion in the training material. The coding tasks from the assessments were used to illustrate the types of contributory factors that should be identified from reports, as well as the associated categories from the classification scheme.

7.5 Conclusions and Next Steps

If a contributory factor classification scheme is going to be implemented within an incident reporting scheme, it is essential that it is reliable and valid. As this chapter illustrates, this is by no means an easy task. This chapter described a step-by-step process for testing and refining contributory factor classification schemes. The example study presented clearly illustrates the challenges associated with designing a valid, reliable, *and* useful classification scheme. It has also highlighted some of the implications for the design of the UPLOADS software tool and associated training materials. The following chapter describes the subsequent design of these components of UPLOADS.

Further Reading

A detailed description of the procedure and materials used to conduct the final assessment of the UPLOADS classification scheme is presented in this paper: Goode, N., Salmon, P. M., Taylor, N. Z., Lenné, M. G., & Finch, C. F. (2017). Developing a contributing factor classification scheme for Rasmussen's AcciMap: Reliability and validity evaluation. *Applied Ergonomics*, 64, 14–26.

References

Annett, J. (2002). A note on the validity and reliability of ergonomics methods. *Theoretical Issues in Ergonomics Science, 3*(2), 228–232. doi:10.1080/14639220210124067.

Finch, C. F., Orchard, J. W., Twomey, D. M., Saleem, M. S., Ekegren, C. L., Lloyd, D. G., & Elliott, B. C. (2012). Coding OSICS sports injury diagnoses in epidemiological studies: Does the background of the coder matter? *British Journal of Sports Medicine, 48*(7), 522–556. doi:https://doi.org/10.1136/bjsports-2012-091219.

Goode, N., Salmon, P., Taylor, N., Lenné, M., & Finch, C. (2017a). Developing a contributing factor classification scheme for Rasmussen's AcciMap: Reliability and validity evaluation. *Applied Ergonomics, 64*, 14–26.

Goode, N., Salmon, P. M., Lenné, M., & Finch, C. F. (2014). A test of a systems theory-based incident coding taxonomy for risk managers. In P. Arezes & P. Carvalho (Eds.), *Advances in Safety Management and Human Factors* (Vol. 10 of Advances in Human Factors and Ergonomics 2014, pp. 5098–5108). United States: AHFE Conference 2014.

Goode, N., Salmon, P. M., Taylor, N. Z., Lenné, M. G., & Finch, C. F. (2017b). Developing a contributing factor classification scheme for Rasmussen's AcciMap: Reliability and validity evaluation. *Applied Ergonomics, 64*, 14–26.

Gordon, R., Flin, R., & Mearns, K. (2005). Designing and evaluating a human factors investigation tool (HFIT) for accident analysis. *Safety Science, 43*(3), 147–171.

Lundberg, J., Rollenhagen, C., & Hollnagel, E. (2009). What-you-look-for-is-what-you-find – The consequences of underlying accident models in eight accident investigation manuals. *Safety Science, 47*(10), 1297–1311. doi:http://dx.doi.org/10.1016/j.ssci.2009.01.004.

Neuendorf, K. A. (2016). *The content analysis guidebook*. Los Angeles: Sage.

O'Connor, P. (2008). HFACS with an additional layer of granularity: Validity and utility in accident analysis. *Aviation Space & Environmental Medicine, 79*(6), 599–606. doi:https://doi.org/10.3357/asem.2228.2008.

Olsen, N. S. (2011). Coding ATC incident data using HFACS: Inter-coder consensus. *Safety Science, 49*(10), 1365–1370. doi:http://dx.doi.org/10.1016/j.ssci.2011.05.007.

Olsen, N. S. (2013). Reliability studies of incident coding systems in high hazard industries: A narrative review of study methodology. *Applied Ergonomics, 44*(2), 175–184. doi:http://dx.doi.org/10.1016/j.apergo.2012.06.009.

Rasmussen, J. (1997). Risk management in a dynamic society: A modelling problem. *Safety Science, 27*(2/3), 183–213.

Ross, A. J., Wallace, B., & Davies, J. B. (2004). Technical note: Measurement issues in taxonomic reliability. *Safety Science, 42*(8), 771–778. doi:http://dx.doi.org/10.1016/j.ssci.2003.10.004.

Stanton, N. A. (2016). On the reliability and validity of, and training in, ergonomics methods: A challenge revisited. *Theoretical Issues in Ergonomics Science, 17*(4), 345–353. doi:10.1080/1463922X.2015.1117688.

Stanton, N. A., & Stevenage, S. V. (1998). Learning to predict human error: Issues of acceptability, reliability and validity. *Ergonomics, 41*(11), 1737–1756. doi:10.1080/001401398186162.

Stanton, N. A., & Young, M. S. (1999). What price ergonomics? *Nature, 399*(6733), 197–198.

Stanton, N. A., & Young, M. S. (2003). Giving ergonomics away? The application of ergonomics methods by novices. *Appl Ergon, 34*(5), 479–490. doi:10.1016/s0003-6870(03)00067-x.

Wallace, B., & Ross, A. J. (2006a). Appendix: Carrying out a reliability trial. *Beyond human error: Taxonomies and safety science* (pp. 243–245). Boca Raton, FL: CRC Press.

Wallace, B., & Ross, A. J. (2006b). *Beyond human error: Taxonomies and safety science*. Boca Raton, FL: CRC Press.

8

Designing a Prototype Incident Reporting System

Practitioner Summary

An incident reporting system underpinned by systems thinking must enable the appropriate reporting, collection, analysis, and aggregation of relevant and high-quality data, while addressing the needs and priorities of end users. To ensure that these considerations are taken into account, the initial development of an incident reporting system should start by specifying design requirements. Design requirements are the goals that must be met for a product to be successful. These requirements should then inform the design of a prototype data collection protocol, learning from incidents process, software tool, and training materials. The chapter begins by describing a starting point for identifying design requirements for an incident reporting system underpinned by systems thinking. It then describes the design requirements identified for UPLOADS, and how they were embedded into the design of the prototype incident reporting system.

Practical Challenges

Minimizing the workload associated with reporting, while enabling the collection of detailed information, is challenging. To address this challenge, the incident report form should not include standardized response options that do not directly contribute to understanding incident causation (e.g. background of the injured person). Instead, the form should prioritize the collection of a detailed qualitative description of the incident and contributory factors.

8.1 Introduction

As discussed in Chapter 3, the quality of systems thinking incident analyses are dependent on the appropriate reporting, collection, analysis, and aggregation of relevant and high-quality data. To feed the analysis, the data collection protocol (i.e. reporting forms) must be designed to support the identification and reporting of relevant details. To ensure that appropriate data is reported, collected, and analyzed, the end users responsible for these tasks must be clearly defined in the learning from incidents process. The software tool and training materials must also be designed to support an appropriate understanding of systems thinking, and the contributory factor classification scheme, among the different end users.

This chapter describes how these considerations, along with end users' priorities, informed the design of the prototype components for UPLOADS. Many of the design requirements identified for UPLOADS are relevant to any incident reporting system that is developed based on the guidance presented in this book. The prototype components developed for UPLOADS are presented as examples of the outputs from the design process.

8.2 Design Requirements

Design requirements are the goals that must be met for a product to be successful. They may be specified as high-level abstract statements, or as detailed formal specifications.

The design requirements for an incident reporting system underpinned by systems thinking should be based on:

- The four core principles of the systems thinking approach for incident reporting systems (outlined in Chapter 1);
- Ensuring that the analyses produced using the contributory factor classification scheme are reliable and valid; and
- End-user needs and priorities.

The following section describes how these principles and constraints were translated into 10 core design requirements for UPLOADS.

8.2.1 Design Requirements for UPLOADS

The design requirements that were developed for UPLOADS based on the systems thinking principles are shown in Table 8.1. These design requirements are relevant to any incident reporting system that is developed based on the guidance presented in this book – they focus on ensuring that

TABLE 8.1

Design Requirements Developed for UPLOADS Based on the Core Principles of Systems Thinking

Principle	Design Requirements
Look 'up and out' rather than 'down and in'	1. The incident report form enables users to identify and report data on contributory factors from across the overall work system.
Identify interactions and relationships	2. The incident report form enables users to identify and report data on potential interactions between contributory factors both within, and across, system levels.
Avoid focusing exclusively on failures	3. The incident report form enables users to describe how the normal conditions of work contributed to incidents as well as identifying perceived 'gaps' in defences.
Apply a systems lens	4. The software tool enables users to use the contributory factor classification scheme (see Chapter 6) to analyze the contributory factors and relationships identified from incident reports. 5. The software tool enables users to represent the contributory factors and relationships involved in multiple incidents on the adapted Accimap framework.

appropriate data is collected, and then analyzed and aggregated using the contributory factor classification scheme.

As discussed in Chapter 7, ensuring that incident reports are analyzed consistently and accurately using a contributory factor classification scheme is by no means an easy task. Even if the classification scheme is designed so the categories are mutually exclusive, exhaustive and neutrally framed, analysts still need to receive appropriate training to ensure that it is used as intended. This issue is ignored in many incident reporting systems, where contributory factor classification schemes are presented as checklists on incident reports for use by untrained reporters. To address this issue within UPLOADS, it was decided to minimize the number of people involved in analyzing incident reports in each organization. Only trained analysts, with relevant LOA expertise, would analyze incident reports (*design requirement 6*).

The design requirements that were identified based on the needs and priorities of LOA practitioners were discussed in detail in Chapter 5. The following requirements were prioritized during the development of the prototype because they were especially important to end users:

- Minimization of workload associated with data reporting (*design requirement 7*);
- Tools for analyzing the data, as well as collecting it (*design requirement 8*);
- Data confidentiality and privacy for individuals reporting incidents and organizations contributing to the sector-wide incident database (*design requirement 9*); and
- Trustworthiness of the collected data and resulting analyses (*design requirement 10*).

As these design requirements were based on the needs and priorities of LOA practitioners, they are not necessarily generalizable to other domains; however, they are intended to provide an example for readers wishing to develop their own.

The following sections describe how these design requirements were embedded in the design of the prototype data collection protocol, learning from incidents process, software tool, and training materials for UPLOADS.

8.3 Prototype Data Collection Protocol

As described in Chapter 3, a data collection protocol needs to describe:

1. The purpose of the system;
2. The scope of the system; and
3. The required data collection fields.

The development of the prototype data collection protocol for UPLOADS is described below.

8.3.1 Purpose of UPLOADS

The purpose of UPLOADS identified by our industry partners was to develop appropriate strategies to prevent, and reduce, incidents and injuries during led outdoor activities (LOA), and to ensure a safer outdoor sector for all. To achieve this goal, the purpose of UPLOADS was refined to support an actionable understanding of:

1. The characteristics of the incidents and injuries that can, and do, occur during LOAs;
2. The relative frequency of incidents associated with different types of LOA; and
3. The network of interacting contributory factors involved in incident causation.

8.3.2 Scope of UPLOADS

Based on the purpose of the system, UPLOADS was limited in scope to only collecting incident and exposure data on LOAs. This meant that UPLOADS was not designed to capture details on other types of occupational injuries occurring within LOA providers, such as injuries associated with cleaning accommodation, or maintaining tracks and camp grounds.

Our initial research indicated that end users believed all incidents associated with adverse outcomes should be reported to UPLOADS, regardless of severity, as well as any 'near miss' incidents. LOA specific definitions for adverse outcomes and near misses were developed based on the World Health Organization's (WHO) guidelines for adverse event reporting and learning systems (Leape & Abookire, 2005).

An '**adverse outcome**' was defined as an event resulting in a negative impact on people, equipment or the environment.

A '**near miss**' was defined as a serious error or mishap that had the potential to cause an adverse event but failed to do so because of chance or because it was intercepted.

8.3.3 Required Data Collection Fields for UPLOADS

Two types of data were required to address the purpose of UPLOADS: incident reports and exposure data. In the LOA context exposure data is referred to as participation data.

8.3.3.1 Incident Report Form

The incident report form was split into two sections: General Incident Information (Table 8.2) and Adverse Outcomes (Table 8.3). The starting point for developing these fields was the New Zealand National Incident Database (NZ NID) Outdoor Education/Recreation Incident Report form. Significant changes were made to improve the standardized response options, and align the form with the Australian LOA context, including:

- The standardized response options for injury type and location were changed to reflect the *International Classification of Diseases,* 10th edition (World Health Organization, 1994).
- The terminology was revised to reflect the Australian LOA context (e.g. 'lost day case' was replaced with 'missing/overdue people').
- Based on our analyses of LOA incidents (Chapter 6), data collection fields were removed that did not directly contribute to understanding incident causation (e.g. ethnicity of people involved in the incident).
- Fields were expanded that contained potentially relevant information for understanding incident causation (e.g. 'Number of people involved' was replaced with separate fields reflecting each role).

TABLE 8.2

UPLOADS Data Fields Describing the General Incident Information

Field	Response Options
Title	Free text
Type of incident	Check boxes (Adverse outcome, Near miss)
Actual/Potential Severity	10-point severity scale from the NZ NID[a]
Standard Operating Procedure (SOP) Number	Free text
Course	Free text
Activity associated with the incident	Free text
Staff supervising activity at the time of incident	Free text
Number of people involved in activity	Number of … participants (e.g. students), activity leaders (e.g. instructors, guides), supervisors (e.g. teachers), volunteers (e.g. parents)
Incident reporter	Free text
Was the reporter present during the incident?	Check boxes (Yes, No)
Incident date	Free text
Incident time	Free text
Location	Free text
GPS co-ordinates	Free text
Narrative. Please provide a detailed description of the incident including: a timeline of events; number of participants, instructors, teachers, etc., present; weather; equipment available; the outcome of the incident (e.g. near miss, injuries, illnesses, social/psychological damage, equipment, and property damage); and any factors you consider played a causal role in the incident and the relationships between them	Free text

[a] A higher score indicated a more severe incident outcome.

Two new features were included on the form to support systems thinking analyses, including:

- Prompts within the incident description to promote the reporting of the contributing factors, and the relationships between them (*design requirements 1 and 2*).

- Examples of the types of contributory factors involved in incidents were included at the end of the form to help users consider contributory factors from across the work system (*design requirement 1*). Most of the examples were framed neutrally to enables users to describe how the normal conditions of work contributed to incidents, as well as identifying perceived 'gaps' in defences (*design requirement 3*).

TABLE 8.3

UPLOADS Data Fields Describing Adverse Outcomes

Field	Response Options
Person impacted	Free text
Experience in activity associated with the incident	Check boxes (Unknown, No prior experience, Some prior experience, Extensive prior experience)
Was the incident fatal	Check boxes (Yes, No)
Injury type[a]	Check boxes (Burns and corrosions; Crushing injury; Dislocation, sprain, or strain; Effects of foreign body entering through natural orifice; Fracture; Frostbite; Injury to internal organs; Injury to muscle, fascia, or tendon; Injury to nerves or spinal cord; Open wound; Poisoning by drugs, medicaments, and biological substances; Sequelae of injuries, of poisoning, and of other consequences of external causes; Superficial injury (e.g. abrasion, blister, insect bite); Toxic effects of substances chiefly nonmedicinal as to source; Traumatic amputation; Other and unspecified effects of external causes)
Injury location[b]	Check boxes (Head; Neck; Chest/Thorax; Abdomen; lower back; lumbar spine and pelvis; Shoulder and upper arm; Elbow and forearm; Wrist and hand; Hip and thigh; Knee and lower leg; Ankle and foot; Multiple body regions; Unspecified part of trunk, limb, or body region)
Illness	Check boxes (Abdominal problem; Allergic reaction; Altitude sickness; Asthma; Chest pain; Diarrhea; Eye infection; Food poisoning; Hypothermia; Heat stroke; Menstrual; Non-specific fever; Skin infection; Respiratory; Urinary tract infection; Unknown; Other)
Social/psychological impacts if applicable	Free text
Treatment at the scene of the incident	Free text
Evacuation method	Free text
Hospitalization required?	Check boxes (Yes, No)
Were emergency services called?	Check boxes (Yes, No)
Overdue or missing people	Check boxes (Yes, No); Free text description
Equipment loss/damage	Check boxes (Yes, No); Free text description
Environmental damage	Check boxes (Yes, No); Free text description

[a] Based on *International Classification of Diseases*, 10th edition (ICD-10, World Health World Health Organization, 1994).
[b] A higher score indicated a more severe incident outcome.

Finally, the data collection fields were limited to two pages to minimize workload (*design requirement 7*).

8.3.3.2 Participation Data Collection Fields

Little guidance was available on the collection of participation data for LOAs. In the organized sports literature, injury rates are typically calculated by dividing the number of injuries or cases (numerator) by the level of exposure to the risk of injury, such as the hours or days of participation (denominator) (Dickson, 2012).

Through feedback from our industry partners, it was identified that some LOA providers collected similar information using course records. These records described the number of participants and activities undertaken during a particular program. To build on this existing workflow, it was decided to standardize the collection of course records to include details on the number of participants, the types of LOA that were participated in, the hours of participation in each activity, and the number of supervisors. This information would then allow the calculation of participation hours, participation days, and participant-to-supervisor ratios.

8.4 Prototype Learning from Incidents Process

A learning prototype from incidents process (Figure 8.1) was developed to describe the processes that would be used to collect, analyze, and translate the lessons learned from incidents into action via UPLOADS.

The prototype process involved activity leaders completing incident reports, and submitting them to a safety manager (or a similar role) within the LOA provider. The safety manager would then enter the reports into the UPLOADS software tool, and analyze the reports using the contributory factor classification scheme embedded into the software tool (*design requirement 4*). This was intended to minimize the number of people involved in analyzing incident reports in each organization (*design requirement 6*).

To ensure confidentiality and privacy, individual LOA providers would be responsible for collecting, storing, and analyzing their own incident data. Only de-identified would be sent to the sector-wide National Incident Database (NID) (*design requirement 9*). This decentralized process was also intended to ensure that LOA providers would be able to immediately analyze, aggregate and act upon their own incident data (*design requirement 8*).

Finally, all data contributed to the NID would be merged prior to analysis, and information identifying organizations would be removed to ensure their confidentiality and privacy (*design requirement 9*). The research team would then analyze, and produce reports on the data, to ensure that resulting

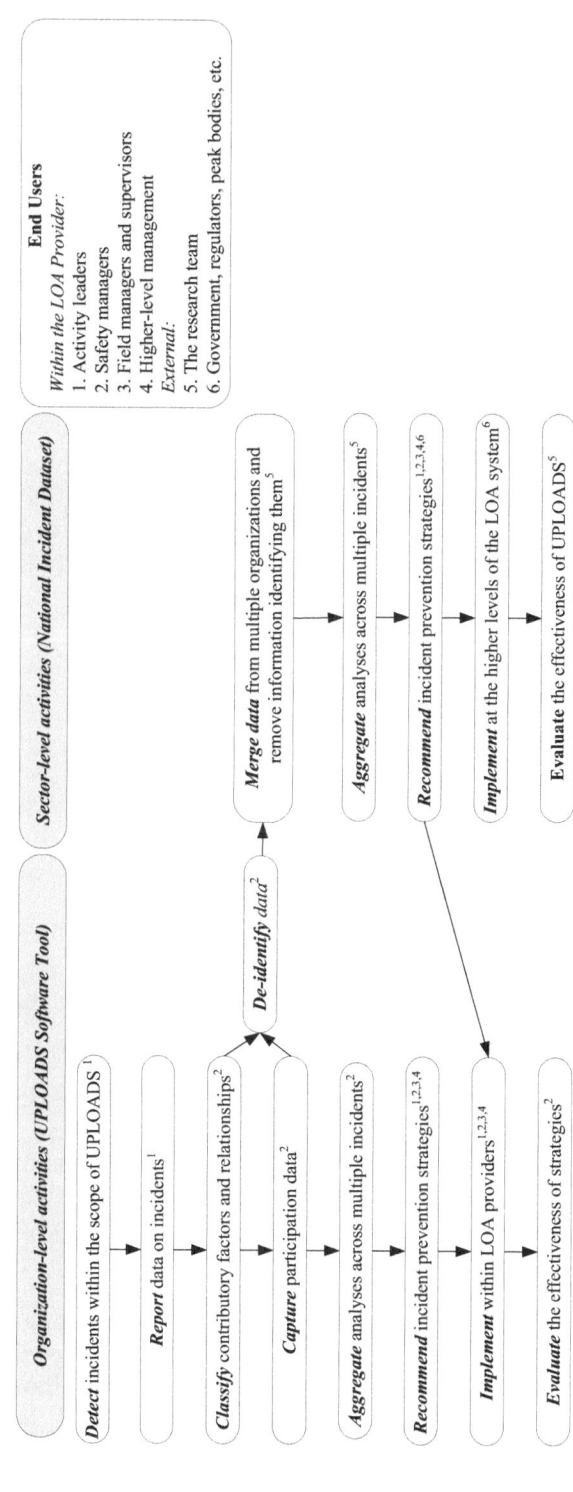

FIGURE 8.1

The prototype learning from incidents processes designed for UPLOADS.

analyses would be trustworthy (*design requirement 10*). It was also intended that the research team would evaluate the effectiveness of UPLOADS over time, by examining the severity and frequency of incidents and injuries during LOA, although this was outside the scope of the original grant.

The process shown in Figure 8.1 then underpinned the design of the prototype UPLOADS software tool.

8.5 Prototype Software Tool

The prototype UPLOADS software tool was developed to support the implementation of the accident analysis method, data collection protocol, and learning from incidents process within LOA providers.

The prototype UPLOADS software tool was intended to provide a comprehensive system for the collection, storage, and analysis of incident and participation data from a systems thinking perspective. The tool consisted of a series of linked databases designed to collect following information:

- Incident reports;
- Staff records;
- Participant/teacher/volunteer records; and
- Course records.

The software tool included the following functions to support the analysis of incident data (*design requirement 8*):

- Exporting of data to spreadsheets for statistical analysis;
- Analysis of incident reports using the contributing factor classification scheme; and
- Production of aggregate Accimaps of analyzed incident data (*design requirement 5*).

In addition, the tool included a function to automatically de-identify data (e.g. names of individuals removed) before submission to the sector-wide NID.

After each incident report form had been entered into the software tool, users were prompted to analyze the incident reports using the prototype UPLOADS contributory factor classification scheme (*design requirement 4*). This involved two steps:

Step 1: The user was prompted to classify the contributing factors using a checklist, as shown in Figure 8.2. For each category that was selected, the user was prompted to enter a description of the specific

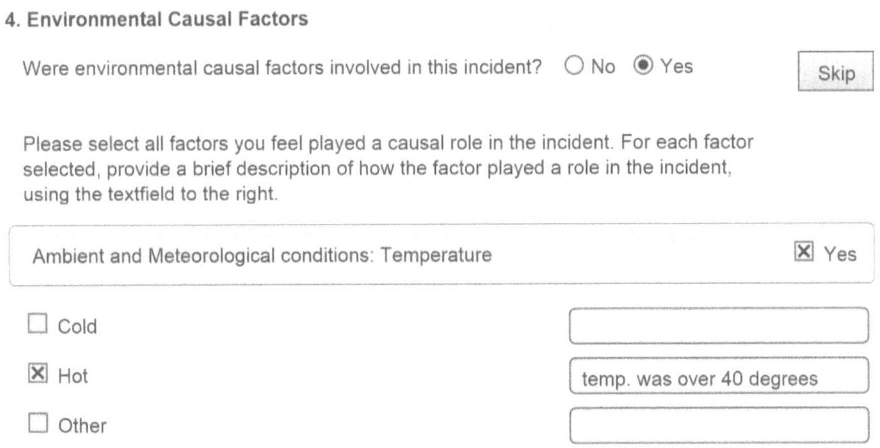

4. Environmental Causal Factors

Were environmental causal factors involved in this incident?   ○ No   ◉ Yes

[Skip]

Please select all factors you feel played a causal role in the incident. For each factor selected, provide a brief description of how the factor played a role in the incident, using the textfield to the right.

| Ambient and Meteorological conditions: Temperature | ☒ Yes |

☐ Cold

☒ Hot

☐ Other

temp. was over 40 degrees

FIGURE 8.2
Example of the UPLOADS prototype software tool user interface for identifying and classifying the factors identified from incident reports.

contributory factor involved in the incident. This was intended to support the interpretation of the aggregated data, and allow the research team to evaluate the validity of classification scheme when used in practice.

Step 2: The user was prompted to identify relationships between any of the contributing factors that had been identified from the report, as shown in Figure 8.3. To classify each relationship, a pair of contributory factors was selected from the drop-down menus. The user was then prompted to enter a description of the specific relationship involved in the report. Again, this was intended to support later validation.

Once incident reports were analyzed, the user was able to create aggregate analyses of multiple incident reports within the software tool, represented on the adapted Accimap framework (*design requirement 5*). Aggregate analyses could be created based on all records in the database or a subset (e.g. only incidents involving a specific activity, or only incidents associated with injuries). Once the desired incident records had been selected, the software tool produced an overall Accimap that produced a graphical summary of the frequency of the contributory factors and relationships identified in the reports. A text file accompanied the Accimap with a table summarizing the descriptions of the contributory factors and relationships that had been identified from the reports.

13. Relationships between factors

Use this section to record relationships between the causal factors you have identified. For example, "inadequate equipment" might be related to "known/planned inappropriate operations" or "poor planning".

If any of the previously selected factors were **related**, then:

1. Select the first factor
2. Select the related factor
3. Briefly describe how you think these factors were related

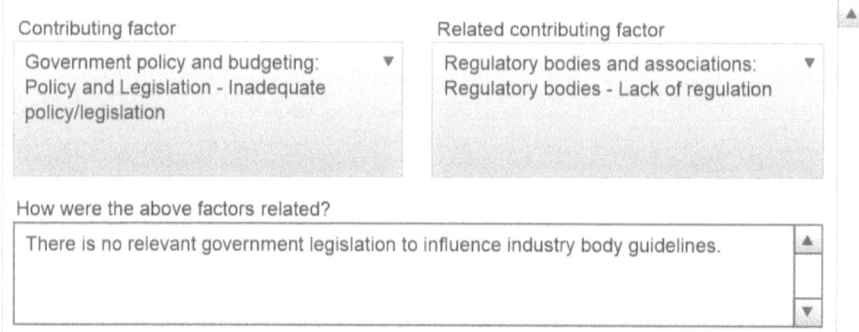

FIGURE 8.3
The UPLOADS prototype software tool user interface for identifying and classifying the relationships identified from incident reports.

8.6 Prototype Training Material

The following training material was developed for the person responsible for implementing UPLOADS within the LOA provider (i.e. safety managers or similar role):

1. A guidebook describing the theoretical approach, and design of UPLOADS and the contributory factor classification scheme;
2. A manual describing how to use the software tool; and
3. Training videos illustrating how to perform each task within the software tool.

For incident reporters (activity leaders), a PowerPoint presentation and script (see Section 8.6.1) was developed describing the systems thinking approach. This primarily focused on appropriately completing the incident report. The presentation was intended to be delivered by safety managers to encourage users to report data regarding potential contributory factors and relationships within, and across, system levels (*design requirements 1 and 2*).

8.6.1 Script for Presentation to Activity Leaders: How to Write a Useful Incident Description

What we can learn from UPLOADS is dependent on the quality of the data that we collect. To fully understand the factors that contribute to incidents, we need to gather as much detail as possible about the circumstances leading up, during, and after the incident.

A good incident description will include a timeline of events that addresses the following questions:

1. Prior to the incident: Are there things that happened prior to the incident itself that you think influenced behavior in a way that enabled the incident to happen? For example:
   - Did any events on the day contribute to the incident?
   - What preparation or planning was undertaken to support the activity?
   - Was this type of incident predicted in training or planning for the activity?
   - Did other similar incidents occur prior to the one being reported?
   - Were there flaws with the training programs, procedures, risk management systems?
   - Were activity programs well designed?
   - Any other details you feel are relevant to the situation?

2. At the time of the incident:
   - What activity was being undertaken?
   - How many people were present (i.e. instructors, participants, teachers, volunteers, others), and who was participating in the activity?
   - What was the weather like?
   - What equipment was being used?
   - Where adequate resources (equipment/staff) available to support the activity?
   - Was adequate information available to support the activity (e.g. weather reports, maps, information on participant allergies and illnesses)?
   - Were there any constraints that shaped how the activity proceeded (e.g. equipment and staff shortages)?
   - Were appropriate communications taking place between activity centre staff?
   - Any other details you feel are relevant to the situation?

3. After the incident:
   - What was the outcome of the incident or why was it considered a near miss?
   - What treatment was provided at the scene?
   - Was evacuation required? How did evacuation occur?
   - Did treatment/evacuation run smoothly?
   - Were adequate resources available for treatment/evacuation?
   - Any other details you feel are relevant to the situation?
4. Sum up the description by suggesting what contributing factors you feel played a role in the incident. If possible, you should also make suggestions on how to prevent future, similar, incidents considering:
   - What would have helped you understand the situation better?
   - Would any specific training, experience, knowledge, procedures or cooperation with others have helped?
   - If a key feature of the situation was different, what would you have done differently?
   - Could clearer guidance have helped you make a better decision?

8.7 Summary and Next Steps

In summary, 10 core design requirements were identified for the design of UPLOADS. These requirements were based on the principles of systems thinking, the identified limitations of the UPLOADS contributory factor classification scheme, and end-user needs and priorities. The design requirements then informed the design of the prototype components for UPLOADS. Each design requirement was clearly reflected in specific aspects of the components.

While the design process should attempt to address the needs and priorities of end users, this does not necessarily ensure that the end product is practical and usable. Following the design of the prototypes, multiple rounds of feedback and testing with end users is usually required to achieve this goal. The following chapter provides guidance on evaluating the usability of an incident reporting system, and refining the system based on the findings.

References

Davidson, G. (2005). *Incident Severity Scale. Adapted and expanded from the Accident Frequency Severity Chart (Priest, 1996).* Retrieved from http://www.incidentre port.org.nz/resources/Severity_Scale.pdf

Dickson, T. J. (2012). Learning from injury surveillance and incident analysis. In T. J. Dickson & T. Gray (Eds.), *Risk Management in the Outdoors: A Whole-of-Organization Approach for Education, Sport and Recreation* (pp. 204–230). Cambridge University Press: Cambridge, England.

Leape, L. L., & Abookire, S. (2005). *WHO Draft Guidelines for Adverse Event Reporting and Learning Systems: From Information to Action.* World Health Organization.

World Health Organization (1994). *International Classification of Diseases,* 10th edition (ICD-10). Retrieved from http://apps.who.int/classifications/icd10/browse /2010/en

9

Evaluating Usability

Practitioner Summary

Usability is the effectiveness, efficiency, and satisfaction with which specified end users can achieve goals in a particular environment. This chapter provides step-by-step guidance on designing and conducting a summative usability assessment to evaluate an incident reporting system. This type of assessment should be undertaken to test, and refine, the design of an incident reporting system prior to implementation. The application of the process is illustrated by describing an evaluation of the UPLOADS prototype. The chapter concludes with a summary of the strategies used to address the problems identified with the UPLOADS prototype, and the implications for the design of any incident reporting system.

Practical Challenges

Summative usability assessments typically involve an evaluation of all major tasks and functionalities with a given system. If the system involves different types of end users, then multiple assessments need to be conducted to identify the usability problems relevant to their required tasks. Although this may be challenging within the limited constraints of project budgets, it is important to remember that it is costlier to fix usability issues after an incident reporting system is implemented.

9.1 Introduction

Involving end users in the design of an incident reporting system is critical to ensure that the system is useable in practice. Early in the development process, informal feedback from end users should be gathered to assess whether proposed designs fit the intended context of application, and to identify potential usability problems. Once the prototypes are at a stage where they are ready to be deployed in an organization, formal assessment processes should be used to evaluate the usability of the system to identify areas for improvement.

This chapter presents a step-by-step guide to conducting a summative usability assessment of an incident reporting systems. One of the assessments undertaken to evaluate the prototype UPLOADS is presented as an example of conducting, reporting, and interpreting the findings from such an evaluation.

9.2 What Is Usability?

The International Standards Organization (ISO) defines usability as the *effectiveness, efficiency,* and *satisfaction* with which *specified end users* are able to achieve *goals* in a particular *environment* (ISO 9241-11, 1998).

- *Effectiveness* is the ability to complete specified tasks, and the success of performing those tasks.
- *Efficiency* is the time required to complete tasks.
- *Satisfaction* is end users' subjective comfort and acceptability of use.

Usability is not a fixed characteristic of a tool or piece of software. It is relative to end users' prior experiences, expectations and attitudes, and the particular tasks they must perform in a given environment (Baber, 2002; Shackel, 2009). The interactions between the end user, tool, task, and environment must be captured to adequately evaluate *effectiveness, efficiency,* and *satisfaction* in a usability assessment (Shackel, 2009).

9.3 Step-by-Step Guide to Evaluating Usability

This section describes how to undertake a summative usability assessment to evaluate an incident reporting system. Summative usability assessments are usually undertaken late in the development cycle (Dumas & Redish, 1999), and typically involve an evaluation of all major tasks and functionalities with a given system. If the system involves different types of end users, then multiple assessments will need to be conducted to identify the usability problems relevant to their required tasks.

TABLE 9.1

End Users, Roles, and Tasks of End Users within UPLOADS Prototype

End User	Tasks
Incident reporter	• Report incidents

Safety manager	• Collect and store:
	  • Participation data
	  • Demographic information about staff members
	  • Demographic information about participants
	  • Incident reports
	• Analyze incident reports using the UPLOADS contributory factor classification scheme
	• Export data for further analysis
	• Produce Accimaps of aggregate incidents

9.3.1 Step 1: Define the Scope of the Assessment

The first step is to define the *end users* and *tasks* that will be included in the assessment. As discussed in Chapter 3, for any given incident reporting system, there are likely to be several different types of end users with different tasks.

As an example, the end users and their tasks within the UPLOADS Prototype are shown in Table 9.1. The usability assessment described in this chapter focuses on the safety manager and their tasks.

9.3.2 Step 2: Identify and Recruit Participants

The second step is identifying and recruiting people to participate in the assessment. A minimum of five to ten participants is considered to be sufficient for identifying a reasonable range of usability problems with a particular system (Dumas & Redish, 1999; Faulkner, 2003).

It is essential to recruit participants who will ultimately perform the tasks in practice (i.e. safety managers or staff in a similar role in the case of the UPLOADS prototype). The recruitment criteria should specify whether participants should have specific skills or background (e.g. safety management), familiarity with the assessment tasks (e.g. novice or familiar end users of the software tool), and experience performing certain tasks (e.g. analysis of contributory factors). These characteristics influence the type of usability problems that will be identified during the assessment.

In addition to end users, it is also useful to include participants with other relevant expertise, such as human factors or data specialists, as they may be able to anticipate additional issues that may impact upon the usability and data quality of the system.

9.3.3 Step 3: Design the Assessment Scenarios

A typical usability assessment involves participants performing a set of pre-determined scenarios using the system. The scenarios should be written in terms of the goal the end user wants to achieve within the system. For example: 'You have been given a paper incident report form by a staff member, and you want to enter it into the UPLOADS software tool.'

Ideally, the scenarios should cover all the tasks relevant to the end users included in the assessment; however, the assessment should run for no more than 1 hour. Typically, 8 to 10 scenarios can be completed in this time.

9.3.4 Step 4: Select Metrics to Use in the Assessment

The next step involves selecting the metrics that will be used to measure the dimensions of usability (i.e. effectiveness, efficiency, and satisfaction) during the assessment.

Effectiveness is usually evaluated using task completion and task success rates. Task completion reflects the percentage or number of scenarios that participants complete correctly with the system (Bailey, 1993; Nielsen, 2001). For example, if the desired goal is to enter a paper-based incident report form into a database, then accuracy could be specified by correctly entering the information into the respective section of the database. Task success reflects the degree to which the tasks have been solved successfully (Macleod & Bowden, 1997). Task success can be scored on a three-point Likert scale (e.g. with ease, with difficulty, failed to complete), or a simple 'Yes or No' response option. Task success can be scored either by an observer, or the participant themselves following each task.

To measure *efficiency*, the expended time in minutes can be recorded for each successfully completed task. This provides a general indication of participants' workload, by trading off the quantity and quality of output against time (Bailey, 1993; Bevan, 1995).

Satisfaction can be measured by obtaining ratings of participants' overall emotional response to the system (e.g. whether the participant thought the software was easy to use, and feels confident using it), their attitudes towards the system, or their perceptions about specific aspects of the system design. Several questionnaires have been developed to collect quantitative ratings of emotional response to a system. Examples include the System Usability Scale (SUS; Brooke, 1996), the Software Usability Measurement Inventory (SUMI; available at http://sumi.uxp.ie), and the Questionnaire for End User Interface Satisfaction (QUIS; Chin, Diehl, & Norman, 1988). The SUS is the most commonly used instrument. Instructions for scoring and interpreting the SUS are available here: https://www.usability.gov/how-to-and-tools/methods /system-usability-scale.html.

There are also several checklists available for collecting qualitative data on the perceived weaknesses of a given system, such as Ravden and Johnson's (1989) Human Computer Interaction (HCI) checklist. This checklist covers nine core domains (e.g. visual clarity, consistency, and compatibility), with prompts to provide written feedback when deficits are identified. This checklist is useful for identifying specific problems related to the user interface.

9.3.5 Step 5: Design the Procedure and Materials

Typically, a usability assessment is conducted face-to-face by a facilitator and observer, with a computer in an office environment. Alternatively, remote testing involves conducting the assessment using the participant's own computer and work environment.

We have used both approaches during the development of UPLOADS, and each has strengths and weaknesses. Face-to-face testing allows for a better understanding of a participant's reactions to the incident reporting system, and results in more detailed qualitative feedback. However, it is more difficult to recruit participants for face-to-face testing, and either the researcher or participant must travel to a specific location to complete the assessment. Remote testing is less expensive as it is conducted via email or post, and allows access to a greater pool of participants. The quality of the feedback is dependent on the information recorded by participants, however, and it is not possible to ask follow-up questions when problems arise.

The procedure used to conduct the assessment determines the materials that are required. In face-to-face testing, participants' responses, and the time taken to complete the scenarios can be recorded by an observer. In remote testing, participants need to either record their responses using a response booklet (i.e. using a stop watch, pen, and paper), or special software can be used to record their responses. In general, the following materials are required:

- A computer;
- Training materials;
- Facilitator guide (face-to-face testing) or an instruction booklet (remote testing); and
- An instruction booklet describing the scenarios.

9.3.6 Step 6: Pilot Test

Prior to conducting the assessment, it is important to test the procedure and materials with at least one participant. The pilot test should be conducted at least 1 to 2 weeks prior to the first test session to allow time to resolve any technical issues, clarify the scenarios or other materials.

9.3.7 Step 7: Conduct the Study

Once the assessment has been piloted, the assessment can be conducted with all participants. It is important to stress to participants that the assessment is testing the design of the system, not their performance; they should provide as much feedback as possible about each scenario, and overall, so that it can be incorporated into the design of the incident reporting system.

9.3.8 Step 8: Data Analysis

The data from each participant should be transcribed into a spreadsheet. The quantitative data for each metric should be summarized using descriptive statistics (e.g. the mean, standard deviation, minimum and maximum). The qualitative feedback should be coded into themes describing the usability problems identified during the assessment.

9.3.9 Step 9: Prioritize Changes to the Design of the Incident Reporting System

An assessment will typically result in many suggestions for improving the design of the system. From a practical point of view, changes to the design of the system can be prioritized based on whether they are likely to improve: (a) usability; (b) the resulting data quality; and (c) acceptability to end users.

9.4 Evaluating the Usability of the UPLOADS Prototype

The previous section presented a step-by-step guide to evaluating the usability of an incident reporting system. This section illustrates the application of the process by describing one of the assessments conducted to evaluate the UPLOADS prototype. The usability assessment described in this section was designed to evaluate both the software tool and training materials.

9.4.1 Scope of the Assessment

The scope of the assessment was limited to the tasks performed by safety managers in the UPLOADS Prototype (see Table 9.1).

9.4.2 Participants

To identify usability problems from multiple perspectives, both LOA safety managers and Human Factors researchers were invited to complete the evaluation. The recruitment criteria are shown in Table 9.2. Participants were 22 LOA safety managers and 12 human factors researchers. LOA safety managers had on average 12.7 years of experience working in the sector. Researchers had on average 9 years of experience working in the field of human factors.

TABLE 9.2

Recruitment Criteria for the UPLOADS Prototype Usability Assessment

LOA Safety Managers	Human Factors Researchers
• Senior staff member in a safety-related role • No prior experience in using the UPLOADS prototype	• Experience in applying systems accident causation models and methods • No prior experience in using the UPLOADS prototype

TABLE 9.3

Scenarios Designed to Assess the Tasks Assigned to Safety Managers

Tasks	Scenarios
Collect and store participation data	You want to enter a template to record information about courses run on a regular basis (Enter a template for a course) You want to enter a record for a course run with a specific group of participants (Enter a course record)
Collect and store demographic information about staff members	You want to create a record for a new staff member, Maya Jones (Enter a staff record)
Collect and store demographic information about participants	You want to import a spreadsheet describing the demographic information of participants involved in a course (Import participant records)
Collect and store incident reports	You want to enter a paper-based incident report into the system (Enter an incident report)
Analyze incident reports using the UPLOADS contributory factor classification scheme	You want to code the contributory factors and relationships involved in the incident report (Code contributing factors and relationships)
Export data for further analysis	You want to export the data that you entered into the software tool onto a spreadsheet for further analysis (Export incident data for analysis)
Produce Accimap	You want to produce an aggregate Accimap, representing all the incidents recorded in the database (Generate an aggregate Accimap)

9.4.3 Assessment Scenarios

The assessment scenarios (shown in Table 9.3) were designed to reflect all of the tasks assigned to safety managers within the UPLOADS prototype.

9.4.4 Metrics

The metrics used to measure effectiveness, efficiency and satisfaction are presented in Table 9.4. Ravden and Johnson's (1989) HCI checklist was used to capture qualitative feedback on end users' satisfaction with the user interface.

TABLE 9.4

Metrics Used to Measure Effectiveness, Efficiency, and Satisfaction in the Assessment

Metric	Questions for Each Scenario
Successful task completion	Did you complete the task successfully? (Options: Yes or No)
Minutes to complete task	How long did you take to complete the task? (Minutes, Seconds)
Satisfaction	Do all the fields make sense?
	Are there any additional fields that should be added?
	Do you have any suggestions or comments for improving the performance of this task?

9.4.5 Procedure and Materials

The assessment was conducted remotely to assist in the recruitment of relevant participants. Participants were provided with an instruction booklet, a response booklet, the prototype software tool, and all associated training material on a USB stick and as hard copies via mail.

The instruction booklet included instructions for installing the prototype software onto the participants own computer, as well as the scenarios for participants to complete (described in Table 9.3). The booklet included data to enter into the prototype software tool for each scenario. Participants were instructed to start by reading through the training materials before beginning the scenarios. Using the response booklet, they were asked to record the time taken to complete each scenario, whether they were able to complete it successfully, as well as any problem encountered or suggestions for improving the system. Following the scenarios, the response booklet then presented Ravden and Johnson's (1989) HCI checklist. Once complete, participants returned the response booklet via mail.

9.4.6 Data Analysis

The data from the booklets were transcribed into a spreadsheet. Descriptive statistics were calculated for successful task completion, and minutes to complete task. The qualitative feedback from each scenario, and the HCI checklist were coded into the themes.

9.4.7 Results

The percentage of LOA practitioners and HF researchers who completed each task successfully, and the time required to complete each task, are shown in Table 9.5.

The feedback provided by LOA practitioners and HF researchers was highly similar. The key issues identified for each task are summarized in Table 9.6.

TABLE 9.5

Task Completion and Time in Minutes Required to Complete Each Task for LOA
Practitioners and HF Researchers

Task Description	Successful Task Completion (% of Participants)		Minutes to Complete Task M (SD)	
	LOA	HF	LOA	HF
Enter a template for a course	95%	83%	1.5 (0.8)	1.3 (0.5)
Enter a record for a staff member	86%	100%	4.3 (1.6)	3.6 (0.9)
Enter a record for a participant	82%	83%	2.8 (1.1)	2.4 (0.8)
Import existing staff records into the database	86%	83%	1.4 (1.1)	1.3 (.8)
Enter a course record	86%	83%	3.4 (1.8)	3.21 (1.7)
Enter an incident report	86%	83%	9.2 (4.7)	7.69 (1.7)
Code contributing factors and relationships	81%	75%	16.9 (9.4)	11.86 (8.6)
Export incident data for analysis	77%	58%	1.6 (0.9)	1.3 (0.52)
Generate an aggregate Accimap	41%	25%	4.2 (4.2)	3.4 (1.4)

9.4.8 Summary of Usability Problems and Resulting Changes

A summary of the key usability problems identified through the assessment,
the resulting changes implemented in the final version of UPLOADS, and
the general implications for the design of incident reporting systems are presented in Table 9.7.

9.5 Summary and Next Steps

This chapter described a step-by-step process for designing and conducting
a summative usability assessment of an incident reporting system. This is
intended to provide a guide for readers wishing to evaluate their own incident reporting system. The evaluation of the prototype UPLOADS allowed
us to identify several usability problems, as well as strategies to address
them. As discussed, many of these strategies are relevant to the design of any
incident reporting system underpinned by system thinking. The next step
in the development process involves conducting an implementation trial to
evaluate the quality of the data collected in practice over an extended period
of time.

TABLE 9.6

Key Issues Identified from Qualitative Feedback Provided by Both LOA Practitioners and HF Researchers

Task Description	Key Issues
Enter a template for a course	• No save 'record' function. • Specify numbers of staff in different roles. • Easy for basic activities, but difficult for camps with many different activities. • Too difficult for large organizations to enter every program. • Confusions calculating participant numbers.
Enter a record for a staff member	• Limitations entering dates in correct format. • Descriptions of roles differ by organizations. • Not required – we already have a staff database.
Enter a record for a participant	• Clarify 'behavioral' issues. • Dropdowns for country are too long. • Role of emergency contact. • Add additional fields (e.g. extra emergency contacts, allergies, swimming ability, height, weight). • We already have a database for participant medical history. • We won't know the answer to many of these questions around medical history.
Import existing staff records into the database	• Import file function hard to find. • Import folder directory hard to find.
Enter a course record	• Limitations entering dates in correct format. • Difficulty adding participant to courses, during to sorting by person ID. • Could not find participant in drop down list. • Very time consuming to complete this section.
Enter an incident report	• Injury type and injury location not in alphabetical order. • Validation of fields makes data entry difficult. • Match paper incident report and database better. • Too many fields. • Put injury types into alphabetical order. • Evacuation method missing some options. • Could not copy and paste from paper form. • What if person has more than one injury? • Organizational response field not clear. • Inability to maximise screen size or scroll makes data entry difficult.
Code contributing factors and relationships	• Presentation as a list is difficult to scroll through. • The list of factors is too long. • Too many overlapping factors.
Export incident data for analysis	• Need feedback that export was successful. • Could not find export directory.
Generate an aggregate Accimap	• Java plugin did not work. • Accimap hard to interpret due to large numbers of factors.

TABLE 9.7

Summary of Usability Problems, the Changes Implemented in UPLOADS, and the General Implications for the Design of Future Incident Reporting Systems

Usability Problems Identified	Changes Implemented in UPLOADS	General Implications
Course template and course records perceived as not required for UPLOADS.	Participation data collection approach redesigned to collect the minimum amount of information required (i.e. number of participants involved in each activity per month, and the total number of days).	Exposure data should be collected at a summary level (e.g. total number of hours worked, patient days), rather than requiring raw data to be entered into the system.
Staff and participant records perceived as not required.	Demographic information on impacted persons integrated into the incident report form, and fields limited to name, age, role, and experience in LOA.	The incident reporting system should collect the minimum amount of information required to draw valid conclusions from the data.
Incident report has too many free text response fields which makes data entry too long.	Standardized response options were developed to capture details on incident characteristics. Soft-copy of the incident report form provided so that information could be copied into the database.	Standardized response options should be utilized to collect as much of the incident data as possible. Minimize duplicated data entry.
Time required to clasify the contributing factors and relationships perceived as too long.	The user interface for classifying contributory factors and relationships was redesigned with drop-down boxes so that relevant codes could more quickly be selected.	The user interface for classifying contributory factors and relationships should minimize the need to search for relevant categories.
Accimap failed to load due to availability of plug-in.	Software changed to eliminate the need for the plug-in.	The software tool should be designed to be compatible with multiple operating systems.
Accimap difficult to interpret.	Further guidance was included in the training material on the interpretation of Accimaps and incorporated into the software tool.	Guidance on interpreting analyses produced by incident reporting systems should be incorporated into the user interface.

References

Baber, C. (2002). Subjective evaluation of usability. *Ergonomics, 45*(14), 1021–1025. doi:10.1080/00140130210166807

Brooke, J. (1996). SUS-A quick and dirty usability scale. *Usability Evaluation in Industry, 189*(194), 4–7.

Chin, P. J., Diehl, A., & Norman, K. (1988). *Development of a tool measuring end user satisfaction of the humam-computer interface.* 19th March 2018, Retrieved from http://lap.umd.edu/quis/publications/chin1988.pdf

Dumas, J. S., & Redish, J. (1999). *A practical guide to usability testing.* Intellect books.

Faulkner, L. (2003). Beyond the five-end user assumption: Benefits of increased sample sizes in usability testing. *Behavior Research Methods, 35*(3), 379–383.

Ravden, S., & Johnson, G. (1989). *Evaluating usability of human-computer interfaces: A practical method.* Halsted Press, New York.

Shackel, B. (2009). Usability – Context, framework, definition, design and evaluation. *Interacting with Computers, 21*(339–346).

10

Evaluating Data Quality

Practitioner Summary

There are five characteristics that are relevant to assessing the quality of data collected via an incident reporting system: data completeness, positive predictive value, sensitivity, specificity, and representativeness. This chapter provides step-by-step guidance on undertaking an implementation trial to evaluate these characteristics. The process described in this chapter can be used for small-scale implementation trials prior to full deployment, and to evaluate incident reporting systems that have already been deployed within an organization. The application of the process is illustrated by describing a 6-month trial of the UPLOADS prototype. The chapter concludes with a summary of the strategies used to address the data quality problems identified from the trial, and the implications for the design of any incident reporting system.

Practical Challenges

Data quality is closely tied to the uptake of an incident reporting system within an organization. Significant planning and effort is required, prior to implementation, to ensure that an incident reporting system is actually used in practice to consistently report, and analyze, incidents.

10.1 Introduction

The development of appropriate incident prevention strategies is reliant on high-quality data about the relative frequency of incidents, and the network of contributory factors involved in them. The data collected must be reliable, valid, representative, and recorded continually over time (German et al., 2001).

Achieving and maintaining such standards is a persistent challenge faced by those who implement, and administer, incident reporting systems (Ekegren, Gabbe, & Finch, 2015).

This chapter presents step-by-step guidance on undertaking an implementation trial to evaluate the data quality of an incident reporting system. The implementation trial undertaken to evaluate the prototype UPLOADS is used to practically illustrate the process.

10.2 What Is Data Quality?

Data quality refers to the completeness and validity of recorded data (German et al., 2001). There are five characteristics that are relevant to assessing data quality in an incident reporting system: *data completeness, positive predictive value, sensitivity, specificity* and *representativeness* (see Table 10.1). These characteristics provide important information about whether the data, and resulting analyses, are accurate and valid reflections of the frequency and causes of incidents within the specific context.

It is important to note that data quality is closely tied to: (a) the usability of a system; and (b) the strategies used to ensure that incidents are consistent and accurately reported and analyzed. If an incident reporting system is hard to use, or if it requires a significant amount of time to complete an incident report, then end users will be unlikely to submit incident reports. Similarly, sufficient time, training, and resources must be provided during implementation to enable end users to consistently and accurately report and analyze incidents. Strategies to improve data quality should focus on both addressing problems with usability, and barriers to effective implementation.

TABLE 10.1

Characteristics of Data Quality in an Incident Reporting System

Characteristic	Definition
Data completeness	A consistent amount of data is provided about every reported incident.
Positive predictive value	Incident reports provide an accurate description of the incident.
Sensitivity	All relevant incidents that occur are reported.
Specificity	No irrelevant incidents are reported.
Representativeness	The incident rates accurately represent how frequently incidents (as defined in the scope) are occurring over time relative to the frequency of exposure.

10.3 Step-by-Step Guide to Undertaking an Implementation Trial

This section describes how to undertake an implementation trial to evaluate the data quality of an incident reporting system. It is good practice to conduct a small-scale trial prior to the full-scale implementation of an incident reporting system, and then periodically evaluate data quality throughout the life cycle of a system. The following process can be used for both types of evaluations.

10.3.1 Step 1: Specify the Aims of the Trial

The first step is specifying the aims of the trial, as it may not be feasible to evaluate all of the characteristics of data quality within a single, small-scale, implementation trial. At a minimum, data completeness and specificity should be evaluated prior full-scale implementation, as they provide information about whether the processes for collecting and analyzing data work in practice.

10.3.2 Step 2: Develop the Evaluation Criteria

The next step is developing criteria for evaluating data quality based on the aims of the trial. Some recommended criteria are presented in Table 10.2. Readers are encouraged to develop additional criteria based on the design of their incident reporting system, and the data available to them.

The evaluation criteria determines the types of data (in addition to the data collected via the incident reporting system) that need to be collected during the trial. For example, evaluating whether the proportion of reported incidents is equal to the actual number of incidents (i.e. sensitivity) requires access to data such as direct observations or case reviews.

10.3.3 Step 3: Specify the Scope of the Trial

The next step is specifying the scope of trial, by describing: (a) who will use the incident reporting system; and (b) how long the trial will run. For example, a small-scale trial of an internal incident reporting system might involve all staff members within a particular department using the system for 6 months; however, an appropriate scope depends on the number of potential reporters, and the frequency of relevant incidents. The scope of the trial needs to be sufficient to ensure that incidents actually occur during the specified timeframe (and so can be reported), and a reasonable number of end users (e.g. relatively to the size of organization) might utilize the incident reporting system.

TABLE 10.2

Recommended Criteria for Evaluating the Data Quality of Incident Reporting Systems Underpinned by Systems Thinking

Characteristic	Definition	Evaluation Criteria
Data completeness	A consistent amount of data is provided about every reported incident.	All fields on each incident report form are completed. All reported incidents are analyzed using the contributory factor classification scheme.
Positive predictive value	Incident reports provide an accurate description of the incident.	All incident reports include details on the contributory factors, and the relationships between them. The incident reports identify contributory factors across all levels of the Accimap framework.
Sensitivity	All relevant incidents that occur are reported.	The proportion of reported incidents is equal to the actual number of incidents. The reports reflect the intended scope of the incident reporting system (e.g. injuries, illnesses, near misses).
Specificity	No irrelevant incidents are reported.	All reported incidents fall within the defined scope of the incident reporting system.
Representativeness	The incident rates accurately represent how frequently incidents (as defined in the scope) are occurring over time relative to the frequency of exposure.	The reported exposure data is equal to the actual exposure data. All incident reports are accompanied by relevant exposure data.

10.3.4 Step 4: Design the Implementation Strategy

The next step is designing an implementation strategy to maximize the uptake, and appropriate use, of the incident reporting system among end users during the implementation trial.

The successful implementation of new practices in organizations is known to involve seven core components: (1) staff selection, (2) training, (3) ongoing coaching and consultation, (4) staff performance assessment, (5) decision support data systems, (6) facilitative administration, and (7) systems interventions (Fixsen, Blase, Naoom, & Wallace, 2009). Potential strategies to address these components during the implementation of an incident reporting system are presented in Table 10.3. The implementation strategies relating to each component should be designed to target the different types of end users within the incident reporting system, and address systemic barriers to implementation.

TABLE 10.3

Core Components to Consider in the Design of Implementation Strategies for Incident Reporting Systems

Component	Implementation Strategy Design
Staff selection	• The characteristics and qualifications of the different types of end users involved in the implementation trial are described. • Methods for communicating with end users with those characteristics and qualifications are described.
Training	• The training is tailored to the different types of end users. • The training provides knowledge of the background theory, core components, and explanations for key practices as well as opportunities to practice and receive feedback.
Ongoing coaching and consultation	• A coach is available to provide ongoing training and feedback when the incident reporting system is used in practice.
Staff performance assessment	• Feedback on the appropriate use of the incident reporting system is provided to end users throughout the trial.
Decision support data systems	• Frequent, user-friendly, reports on progress and success are provided to decision-makers to ensure barriers to implementation are addressed at the organizational level.
Facilitative administration	• The policies, procedures, structures and culture of the organization support the implementation of the incident reporting system.
Systems intervention	• Adequate financial, organizational, and other resources are available to support the implementation of the incident reporting system.

10.3.5 Step 5: Conduct the Implementation Trial

The next step is deploying the incident reporting system, along with the implementation strategy. It is important to closely monitor the progress of data collection throughout the trial, and resolve any technical difficulties that are encountered by end users. If no data is collected during the first few months of the trial then new strategies may need to be developed to improve uptake, or the trial may need to be terminated to resolve the barriers to implementation.

10.3.6 Step 6: Data Analysis

The data collected using the incident reporting system should be merged into a spreadsheet for analysis, along with any additional data that has been collected during the trial. The quantitative data addressing the evaluation criteria should be summarized using descriptive statistics (e.g. the mean, standard deviation, minimum, and maximum). The qualitative data addressing the evaluation criteria should be coded into themes, and summarized.

10.3.7 Step 7: Identify Solutions to Address the Data Quality Problems

An implementation trial will usually identify several data quality problems relating to: (a) the design of different components of the incident reporting system (e.g. the questions used to collect data, or the software user interface), and (b) the implementation strategy. It is important, however, not to focus too much on the training aspects of the implementation strategy. While reinforcing the need to report incidents may improve data quality in the short term, resolving problems with the design of the system, and systemic barriers to using it, will result in better data quality over the long term.

10.4 Evaluating the Data Quality of the UPLOADS Prototype

The previous section presented a step-by-step guide to designing an implementation trial to evaluate the data quality of an incident reporting system. This section illustrates the application of this process to the UPLOADS prototype. The prototype was described in detail in Chapter 8.

10.4.1 Aims of the Implementation Trial

The aims of the implementation trial were to evaluate the data completeness, positive predictive value, sensitivity, specificity, and representativeness of the data collected via the UPLOADS prototype when used by LOA providers.

10.4.2 Evaluation Criteria

The criteria used to evaluate the data collected during the implementation trial are shown in Table 10.4. The criteria were developed based on the aims of the trial, and the design of the UPLOADS prototype.

10.4.3 Scope of the Trial

The trial involved LOA providers using the UPLOADS prototype to collect and analyze data for 6 months (June to December 2013). The intention was to recruit at least one LOA provider from each state and territory of Australia, representing a diverse range of organizations.

In total, 15 organizations from across Australia participated in the trial, including 5 commercial enterprises, 5 not-for-profits, 2 schools, 2 registered training organizations, and 1 public sector agency. Organizations ranged in size from sole operators with 1 full-time staff member, to large organizations operating across multiple sites with 150 full-time staff.

10.4.4 Implementation Strategy

The implementation strategy (shown in Table 10.5) primarily focused on recruiting, and maintaining the participation of safety managers during the trial,

TABLE 10.4

Criteria for Evaluating the Data Quality of the Prototype UPLOADS

Characteristic	Evaluation Criteria
Data completeness	All fields on each incident report form are completed.
	All participation records include the total number of: participants, participation hours, and participation days for each activity.
	All reported incidents are analyzed using the contributory factor classification scheme.
Positive predictive value	All incident reports include details on the contributory factors, and the relationships between them.
	The incident reports identify contributory factors across all levels of the Accimap framework.
Sensitivity	The incident reports include adverse outcomes and near misses.
	The incident reports include many incidents with a low rated severity.
Specificity	All reported incidents involve incidents that occur during LOAs.
Representativeness	All incident reports are accompanied by relevant exposure data.
	The organizations participating in the trial provide data to the research team on a monthly basis.

as they provided the primary point of contact for the research team. This was intended to have a flow-on effect to reporters throughout the organization.

10.4.5 Procedure

Organizations and safety managers completed the consent form to participate in the trial. Safety managers completed an online demographics questionnaire, which asked for details on their background, and the organization.

Safety managers were then sent the trial materials via mail, which included soft and hard copies of all documents and the software tool on a USB stick. They were instructed to: (1) provide training to staff members responsible for reporting incidents; (2) read through the manuals and watch the videos prior to installing the software tool; (3) install the software tool on a computer; (4) enter incident reports and participation data into the software tool and analyze the reports; and (5) contribute de-identified data to the research team on a monthly basis. Ongoing coaching and performance feedback for safety managers was provided throughout the trial, as described in Table 10.5. At the end of the trial, all safety managers were sent a survey to gather their feedback on the software tool, and training materials.

During the initial follow-up, many safety managers indicated that the section of the software tool used to record participation data (entering templates, course records, and details for all participants) was an overly burdensome requirement. In response, an excel spreadsheet was provided to collect participation data. The spreadsheet recorded the total number of participants, the total number of participation hours, and the total number of participation days during a calendar month.

TABLE 10.5

Implementation Strategy for the UPLOADS Prototype Trial

Component	Safety Managers	Incident Reporters
Staff selection	Characteristics: senior staff member in a safety-related role. Recruitment: nominated by their organization.	Characteristics: all staff members involved in the delivery of LOA (e.g. activity leaders). Recruitment: informed of the trial by the safety manager.
Training (pre-trial)	Information sheet explaining the purposes and aims of the trial. Two written manuals describing the theoretical approach and design of UPLOADS, and how to use the accident analysis method and software tool. Training videos illustrating how to perform each task within the software tool.	Information sheet explaining the purposes and aims of the trial. PowerPoint presentation describing the systems thinking approach and the types of information they needed to include on incident reports.
Ongoing coaching throughout trial	One-on-one support available from the research team via email and telephone. Followed up via phone two weeks into the trial to confirm that they had received the materials, understood the training materials, and were able to successfully use the software tool. Monthly reminders to submit de-identified data to the research team.	
Staff performance assessment	Feedback provided to confirm that de-identified data was successfully submitted to the research team.	

10.4.6 Results from the Implementation Trial

In total, 183 incidents were reported during the trial, with participation data recorded on 59 activities. The following sections present the results from the trial relevant to the evaluation criteria.

10.4.6.1 Data Completeness: Incident Reports

Although all incident reports (n = 183) included a narrative description of the incident, more than 50% of reports included missing fields. The following fields were not completed in the majority of reports: Standard operating procedure (SOP) number; course; location; GPS coordinates; and organizational response to the incident. This indicated that potentially there was a need to remove some of these data collection fields.

Severity scores were validated by the research team using the information provided in the report. More severe incidents tended to be rated by incident reporters as higher in severity than justified by the information provided in the report. This indicated that the incident severity scale had poor validity.

10.4.6.2 Data Completeness: Participation Data

Only one safety manager used the course records feature to record participation data; all other safety managers used the Excel spreadsheet provided during the trial, which required only summary data. The number of participation hours was not consistently recorded; some safety managers reported it was impossible to calculate participation hours for multi-day activities such as camping. Participation days and participant numbers were completed consistently.

10.4.6.3 Data Completeness: Analysis of Incident Reports

Almost all (152 out of 183) incident reports were analyzed by safety managers using the contributory factor classification scheme provided in the software tool.

10.4.6.4 Positive Predictive Value

Almost all (152 out of 183) incident reports included details on contributory factors, but only 37 incident report included details on relationships between factors.

The incident reports did not include sufficient detail to support the identification of contributory factors across all levels of the Accimap framework. As shown in Figure 10.1, safety managers only identified contributory factors at the lower four levels of the framework.

10.4.6.5 Sensitivity

Overall, the incident reports reflected the intended scope of the prototype UPLOADS. The reports included 159 incidents associated with adverse outcomes and 25 near misses. In addition, as shown in Figure 10.2, many incidents were reported with a low severity rating, in line with the criteria for reporting.

10.4.6.6 Specificity

All incident reports involved incidents that occurred during LOA.

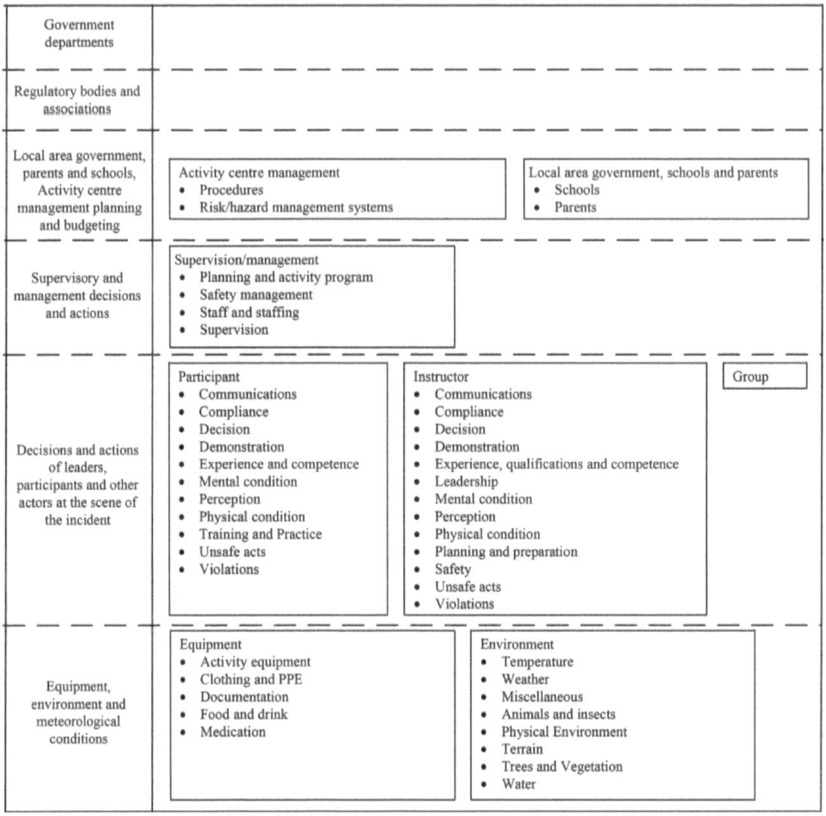

Government departments
Regulatory bodies and associations
Local area government, parents and schools, Activity centre management planning and budgeting
Activity centre management
• Procedures
• Risk/hazard management systems
Local area government, schools and parents
• Schools
• Parents
Supervisory and management decisions and actions
Supervision/management
• Planning and activity program
• Safety management
• Staff and staffing
• Supervision
Decisions and actions of leaders, participants and other actors at the scene of the incident
Participant
• Communications
• Compliance
• Decision
• Demonstration
• Experience and competence
• Mental condition
• Perception
• Physical condition
• Training and Practice
• Unsafe acts
• Violations
Instructor
• Communications
• Compliance
• Decision
• Demonstration
• Experience, qualifications and competence
• Leadership
• Mental condition
• Perception
• Physical condition
• Planning and preparation
• Safety
• Unsafe acts
• Violations
Group
Equipment, environment and meteorological conditions
Equipment
• Activity equipment
• Clothing and PPE
• Documentation
• Food and drink
• Medication
Environment
• Temperature
• Weather
• Miscellaneous
• Animals and insects
• Physical Environment
• Terrain
• Trees and Vegetation
• Water

FIGURE 10.1
Types of contributory factors identified from the incident reports by safety managers during the implementation trial.

35
30
25
20
15
10
5
0
No. of incident reports
1
2
3
4
5
6
Severity ratings

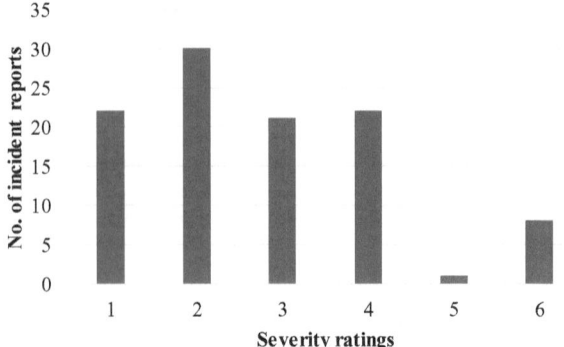

FIGURE 10.2
Severity ratings of incidents associated with adverse outcomes. (Scale: 1–3: Short-term impact; 4–5: Medium impact; 6–7: Major impact; and 8–10: Life-changing effect or death.)

10.4.6.7 Representativeness

All incident reports were accompanied by relevant participation data (i.e. relative to the time of the incident report and activity). The reported incidents involved 31 different activities, and 59 activities were recorded in the participation data (as might be expected). There were many similar activities, which fitted into 16 broad categories; however, many activities (e.g. horseback riding) were only conducted by one organization. This raised concerns about how to appropriately report on incidents associated with these activities at the sector-wide level.

The submission of data to the research team was maintained initially and then dropped sharply over the last 2 months of the trial, as shown in Figures 10.3 and 10.4. Only 6 (out of the 15) safety managers responded to requests for data in every month of the trial, and two did not respond

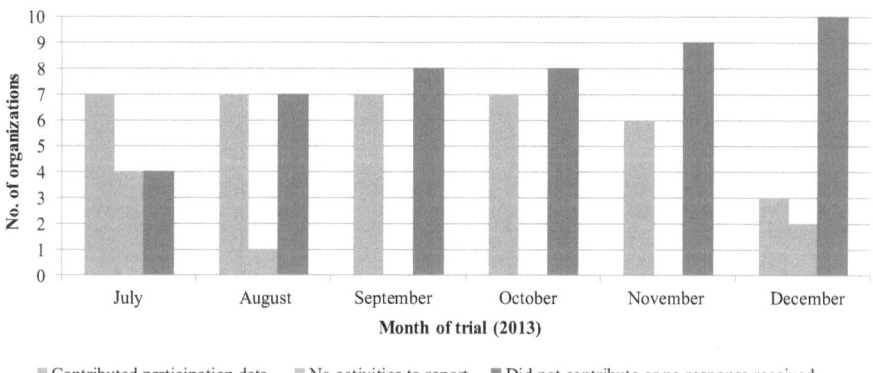

FIGURE 10.3
Number of organizations contributing participation data to UPLOADS in each month of the trial.

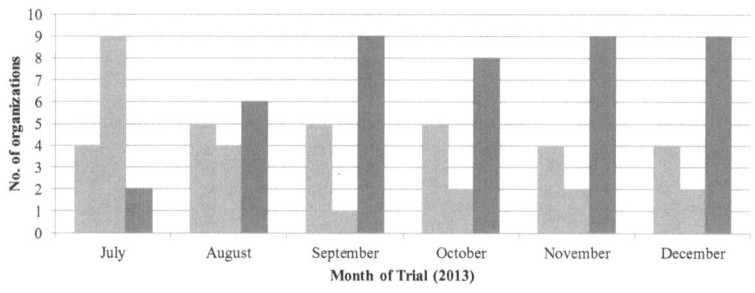

FIGURE 10.4
Number of organizations contributing incident data to UPLOADS in each month of the trial.

to any. Several organizations did not have participation or incident data to report each month as they only provided seasonal LOAs. Similarly, several safety managers reported that they had no incidents to report during the period, despite conducting LOAs.

Safety managers that contributed data infrequently or not at all stated that they: were overwhelmed by their workload and had no time to use the system; had staff shortages; already had an incident reporting system and it was not time efficient to use both that and UPLOADS; and lacked support from management. One safety manager reported that the program was not compatible with their office network.

10.4.6.8 Summary of Problems and Implications

A summary of the data quality problems identified, the changes to the design of UPLOADS, and the implications for the design of incident reporting systems more generally is presented in Table 10.6.

TABLE 10.6

Summary of Data Quality Problems, the Changes Implemented in the Design of UPLOADS, and the Implications for the Design of Incident Reporting Systems

Problems	Changes Implemented in UPLOADS	Implications for Incident Reporting Systems
More than 50% of incident reports included missing fields, which tended to be questions with free text fields.	Standardized response options were developed to describe all incident characteristics.	Standardized response options should be utilized to collect as much of the incident data as possible.
The incident severity scale had poor validity, due to the subjective nature of the criteria.	A simplified version of the original scale was developed, with objective criteria for each severity level (e.g. 5 = requires urgent evacuation and emergency assistance with long term effects on the person impacted).	Objective criteria should be used to define any rating scales used within an incident reporting system.
The participation data collection fields required too much time to complete.	The course template and course records functions were replaced with a section for entering summary information about participation during a single month.	Exposure data should be collected at a summary level (e.g. total number of hours worked, patient days), rather than requiring raw data to be entered into the system.

(Continued)

TABLE 10.6 (CONTINUED)

Summary of Data Quality Problems, the Changes Implemented in the Design of UPLOADS, and the Implications for the Design of Incident Reporting Systems

Problems	Changes Implemented in UPLOADS	Implications for Incident Reporting Systems
The number of participation hours was recorded inconsistently.	The requirement to report on participation hours was removed. Participation numbers and days per activity were used to capture exposure data.	Consider the level of granularity required for exposure data to produce useful results.
Many similar activities were recorded in the participation data.	A classification scheme was developed to standardize the reporting of the types of LOAs involved in incidents and participation data.	Standardized response options should be utilized to collect as much of the incident data as possible.
The descriptions of the incidents did not include sufficient detail to support the identification of contributory factors and relationships across all levels of the Accimap framework.	Prompts were developed to assist in the elicitation of a more detailed incident description. Expansion of the incident reporting process to collect information from Supervisors or Field Managers, in addition to Activity Leaders.	Include prompts in the incident report form to extract information about contributory factors away from the sharp end of the system (e.g. outside of people, equipment, and the environment). The reporting system should possess the capacity for reports to be completed by various reporters (e.g. the ability for a supervisor to add further details to an existing report).
Many activities were only conducted by one organization, raising concerns about privacy and confidentiality.	A protocol was developed specifying the minimum cases required for the research team to report on specific types of incidents and activities.	For sector-wide incident reporting systems, the process for analyzing and reporting on data should be designed to protect the confidentiality and privacy of organizations.
Only 6 out of the 15 organizations responded to requests for data in every month of the trial.	Face-to-face workshops were developed to provide training on UPLOADS. This training was made available prior to the national deployment of UPLOADS, and at regular intervals to maintain participation.	Implementation strategies should be designed in collaboration with end users to address issues associated with uptake.

10.5 Summary and Next Steps

This chapter described a step-by-step process for undertaking an implementation trial, and evaluating the data quality of an incident reporting system. This is intended to provide a guide for readers wishing to evaluate their own incident reporting system. The implementation trial described in this chapter provided important information about barriers to using the UPLOADS prototype in practice over an extended period. The findings from the implementation trial, and the usability assessment (see Chapter 9), led to the development of a revised version of UPLOADS, which is described in the following chapter.

Further Reading

Fixsen et al. (2009) provides further guidance on the development of successful implementation strategies.

References

Ekegren, C. L., Gabbe, B. J., & Finch, C. F. (2015). Injury surveillance in community sport: Can we obtain valid data from sports trainers? *Scandinavian Journal of Medicine & Science in Sports, 25*(3), 315–322. doi:doi:10.1111/sms.12216

Fixsen, D. L., Blase, K. A., Naoom, S. F., & Wallace, F. (2009). Core implementation components. *Research on Social Work Practice, 19*(5), 531–540.

German, R., Lee, L., Horan, J., Milstein, R., Pertowski, C., & Waller, M. (2001). Guidelines Working Group Centers for Disease Control and Prevention (CDC). Updated guidelines for evaluating public health surveillance systems: Recommendations from the Guidelines Working Group. *MMWR Recomm Rep, 50*(RR-13), 1–35.

11

Outputs from the Development Process – UPLOADS

Practitioner Summary

This chapter presents an overview of UPLOADS, as implemented in the Australian outdoor sector in 2014. The chapter describes the system, including:

- Domain-specific accident analysis method with a contributory factor classification scheme;
- Data collection protocol;
- Process for learning from incidents at both an individual organization, and sector-wide, level;
- Software tools to support the implementation of the data collection protocol and accident analysis method within organizations; and
- Training materials.

The chapter describes these components, highlighting the improvements that were made based on the usability and data quality evaluations described in Chapters 9 and 10.

11.1 Introduction

This chapter describes the version of UPLOADS that was implemented in the Australian LOA sector from the 1st June 2014. UPLOADS comprised the following components, which were developed to support a sector-wide systems thinking approach to reporting, analyzing, and preventing LOA incidents:

- A domain-specific accident analysis method based on Accimap, with a contributory factor classification scheme;
- A data collection protocol describing the purpose and scope of the system, what types of incidents should be reported, and what data are collected;
- A process for learning from incidents within individual organizations, and across the sector;
- A software tool to support implementation within LOA providers; and
- Training materials for the person responsible for implementing UPLOADS within their organization (i.e. safety manager), and incident reporters.

The following sections describe these components in detail, highlighting the significant changes that were made as a result of the usability and data quality evaluations (Chapters 9 and 10). While the components were developed specifically for the LOA-context, they could be adapted for a different domain by readers wishing to develop their own incident reporting system.

11.2 Accident Analysis Method

The accident analysis method forms the core of UPLOADS, and it influenced the design of all the other components. The method draws on Rasmussen's (1997) risk management framework and the associated Accimap (Svedung & Rasmussen, 2002). It is underpinned by a contributory factor classification scheme incorporating categories across the five levels of the LOA system, as shown in Figure 11.1. The classification scheme provides a standardized approach to analyzing the qualitative information captured in incident reports, which enables aggregate data to be represented on the accident analysis framework.

The classification scheme has a two-level structure. Level 1 describes the LOA system in terms of the:

- Actors (e.g. activity leaders, activity participants, supervisors, parents);
- Organizations (e.g. state and federal government, regulatory bodies and associations, schools);

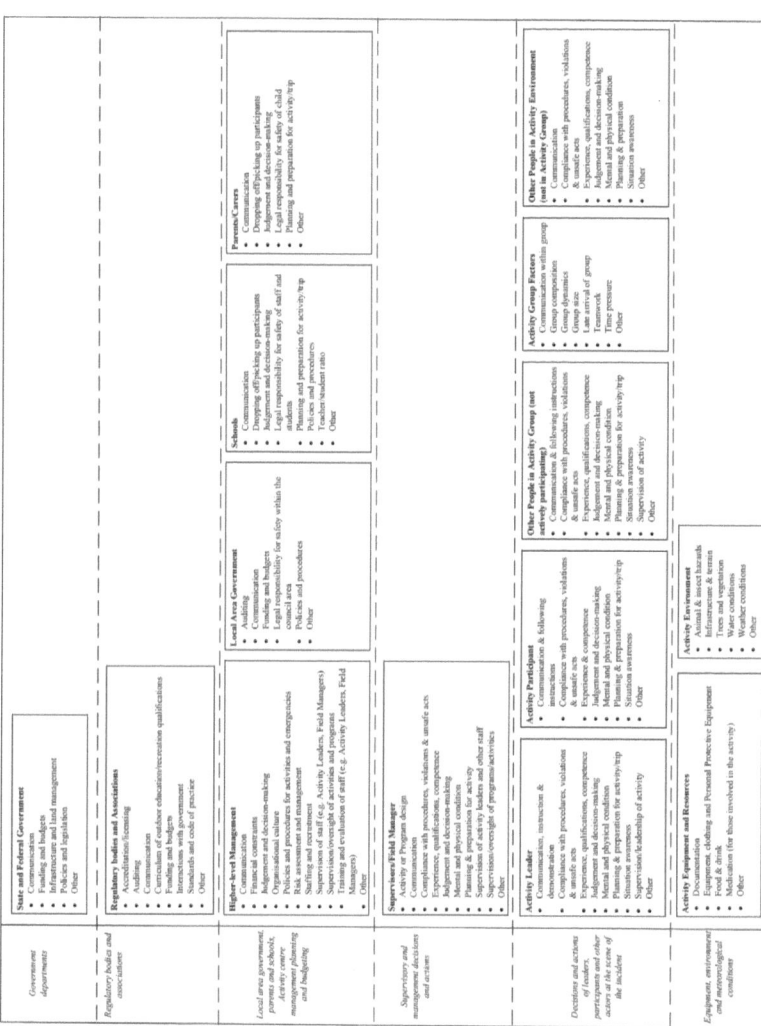

FIGURE 11.1

UPLOADS accident analysis method and contributory factor classification scheme. (From Goode, N., Salmon, P. M., Taylor, N. Z., Lenné, M. G., & Finch, C. F. 2016. *Theoretical Issues in Ergonomics Science*, 17(5–6), 483–506. With permission.)

- Artefacts (e.g. activity equipment); and
- Activity context (e.g. environment and relevant conditions).

Level 2 breaks the Level 1 categories down into specific contributory factors, with a total of 107 categories. As the classification scheme was developed based on incident data from the United Kingdom, New Zealand, and Australia, it is likely to be applicable to international LOA contexts, not just Australia.

11.3 Data Collection Protocol

The data collection protocol describes the purpose and scope of UPLOADS, and the types of data collected.

11.3.1 Purpose of UPLOADS

The purpose of UPLOADS was to support LOA providers and the sector to understand:

1. The characteristics of the incidents and injuries that can, and do, occur during LOAs;
2. The relative frequency of incidents associated with different types of LOA; and
3. The network of interacting contributory factors involved in incident causation.

The long-term goal was to gather, and analyze, incident data to support the development of appropriate prevention strategies to reduce LOA incidents and injuries, to support safer LOAs for all.

11.3.2 Scope of UPLOADS

UPLOADS was designed to collect data about *incidents that occur during LOAs*, involving adverse outcomes or a near miss.

- An **incident** was defined as an event that results in an adverse outcome or a near miss during an LOA.
- An **adverse outcome** was defined as an event resulting in a negative impact. For LOA's negative impacts include any injury, illness, fatality, equipment damage, environmental damage, and social or

psychological impacts occurring during the activity, as well as missing or overdue people returning from the activity.

- A **near miss** was defined as a serious error or mishap that has the potential to cause an adverse event, but fails to do so because of chance or because it is intercepted. For example, during a rock climbing activity an activity leader notices that a participant's carabineer was not locked. If the student had fallen, this may have led to a serious injury.

These definitions were adapted from those provided in the World Health Organization's guidelines for adverse event reporting and learning systems (Leape & Abookire, 2005).

To further distinguish between near misses and adverse outcomes, a LOA-specific incident severity scale (Table 11.1) was developed to rate the actual and potential severity of incidents. Actual severity was based on the response to the event and the outcome for the injured person, while potential severity was based on reporter's subjective perception of the

TABLE 11.1

Incident Severity Rating Scale, with Definitions for Rating Incidents in Terms of Their Actual and Potential Severity

Rating	Incident	Actual Severity Ratings	Potential Severity Ratings
0	No impact	Requires no treatment.	An incident where the potential outcome has a negligible consequence.
1	Minor	Requires localized care (non-evacuation) with short-term effects.	An incident where the potential outcome to risks has a low consequence.
2	Moderate	Requires ongoing care (localized or external, i.e. evacuation or not) with short- to medium-term effects.	An incident where the potential outcome to risks can cause moderate injuries or illnesses.
3	Serious	Requires timely external care (evacuation) with medium to long term effects.	An incident where the potential outcome to risks encountered is such that it may cause major irreversible damage or threaten life.
4	Severe	Requires urgent emergency assistance with long-term effects.	An incident where the potential outcome to risks encountered is certain death.
5	Critical	Requires urgent emergency assistance with serious ongoing long-term effects.	–
6	Unsurvivable	Fatality.	–

worst possible outcome given the scenario. To be classified as a near miss, incidents must be allocated an actual severity rating of '0', indicating no impact. Any incident with an actual severity rating of 1–6 was to be classified as an adverse outcome. The scale is a simplified version of the 10-point incident severity scale (Davidson, 2005) that was used in the prototype UPLOADS. The revised scale was developed in response to problems with the validity of the original scale, identified in the implementation trial (see Chapter 10).

The criteria for reporting incidents to UPLOADS was based on incident severity, in line with the WHO guidelines (Leape & Abookire, 2005). UPLOADS users were instructed to report any:

- Adverse outcome with an actual severity of 1 or greater; and
- Near misses with a potential severity of 2 or greater.

These criteria were established to ensure that the data collected via UPLOADS was representative of all incidents that occur during LOAs, and not biased towards more serious incidents.

The criteria for reporting near misses was developed based on the finding that very few near misses were reported during the implementation trial (Chapter 10). The criteria for reporting near misses (potential to cause moderate injuries or illnesses) was established to encourage greater reporting, based on feedback that there was a perception that only near misses with potentially very serious or life-threatening outcomes should be reported.

11.3.3 Data Collection Fields

UPLOADS supported the collection of two types of data: incident and participation data.

The incident data collection fields fitted into three categories: *Incident Characteristics*, *Adverse Outcomes*, and *Incident Description*. The fields formed the basis of the incident report form that was used by organizations to collect data (see Appendix C).

The *Incident Characteristics* fields, shown in Table 11.2, were designed to capture information about the type and severity of the incident, contextual factors at the time of the incident, and some of the factors that may have mitigated or worsened the incident. The free text fields included in the prototype version of the incident report form were removed, based on the low completion rates identified in the implementation trial (Chapter 11).

The *Adverse Outcomes* fields, shown in Table 11.3, were designed to capture the outcome and immediate response to the incident. These fields were not completed for near misses. An important feature of this section was that the

TABLE 11.2

UPLOADS Data Fields Used to Collect Information about Incident Characteristics

Field	Response Options
Incident reporter name	Free text
Was the reporter present at the incident?	Check boxes (Yes, No)
Date/Time	Free text
State/Territory	Free text
Staff responsible for supervision at the time of the incident	Free text
Type of incident	Check boxes (Adverse outcome, Near miss)
Actual severity rating	Incident Severity Scale (0-6)
Potential severity rating	Incident Severity Scale (0-4)
Activity associated with the incident	Free text
Main goals associated with activity	Free text
Weather at the time of incident	Rain conditions (1 = Fine, 4 = Wet) Temperature (1 = Hot, 4 = Cold) Wind conditions (1 = Calm, 4 = Windy)
Number of people involved in activity	Number of (participants (e.g. students), activity leaders (e.g. instructors, guides), supervisors (e.g. teachers), volunteers (e.g. parents)
Location of incident	Free text
Did the activity leader have relevant qualifications?	Check boxes (Yes, No)

information would be sufficient to validate the actual severity ratings provided by the reporter.

The *Incident Description* fields, shown in Table 11.4, were designed to capture detailed qualitative information about the incident to support the application of the UPLOADS accident analysis method and classification scheme. This included a description of what happened, who was involved, any relevant events prior to the incidents, and contributory factors the reporter and their manager felt played a role in the incident. These fields were presented with the following instructions on the incident report form:

> It is very important that you identify all the factors, and the relationships between them, which may have contributed to the incident you are reporting. To assist you in thinking about the contributory factors involved in your incident, we have provided examples below of factors that have been found to play a role in previous incidents.

The contributory factor classification scheme (shown in Figure 11.1) was then presented on the incident report form to prompt consideration of the range of contributory factors that are potentially involved in LOA incidents.

TABLE 11.3

UPLOADS Data Fields Used to Collect Information about Adverse Outcomes

Field	Response Options
Person impacted	Free text
Experience in activity associated with the incident	Check boxes (Unknown, No prior experience, Some prior experience, Extensive prior experience)
Was the incident fatal?	Check boxes (Yes, No)
Injury type[1]	Check boxes (Burns and corrosions; Crushing injury; Dislocation, sprain, or strain; Effects of foreign body entering through natural orifice; Fracture; Frostbite; Injury to internal organs; Injury to muscle, fascia, or tendon; Injury to nerves or spinal cord; Open wound; Poisoning by drugs, medicaments, and biological substances; Sequelae of injuries, of poisoning, and of other consequences of external causes; Superficial injury (e.g. abrasion, blister, insect bite); Toxic effects of substances chiefly nonmedicinal as to source; Traumatic amputation; Other and unspecified effects of external causes)
Injury location[a]	Check boxes (Head; Neck; Chest/Thorax; Abdomen; lower back, lumbar spine, and pelvis; Shoulder and upper arm; Elbow and forearm; Wrist and hand; Hip and thigh; Knee and lower leg; Ankle and foot; Multiple body regions; Unspecified part of trunk, limb, or body region)
Illness	Check boxes (Abdominal problem; Allergic reaction; Altitude sickness; Asthma; Chest pain; Diarrhea; Eye infection; Food poisoning; Hypothermia; Heat stroke; Menstrual; Non-specific fever; Skin infection; Respiratory; Urinary tract infection; Unknown; Other)
Evacuation method	Check boxes (Boat; Helicopter; Ski patrol stretches; Sled; Stretcher; Snowmobile; Vehicle; Walked out)
Hospitalization required?	Check boxes (Yes, No)
Was the emergency services called?	Check boxes (Yes, No)
Briefly describe the social/psychological impacts on the person described above	Free text
Briefly describe any treatment at the scene of the incident	Free text
Overdue or missing people	Check boxes (Yes, No); Free text description
Equipment loss/damage	Check boxes (Yes, No); Free text description
Environmental damage	Check boxes (Yes, No); Free text description

[a] Based on *International Classification of Diseases*, 10th edition (ICD-10, World Health World Health Organization, 1994).

TABLE 11.4

UPLOADS Data Fields Used to Collect Information about What Happened during the Incident and Contributory Factors

Field	Response Options
Describe the incident in detail, include: who was involved, what happened, when it happened, where it happened, and any equipment involved.	Free text
Describe any relevant events leading up to incident.	Free text
Describe why the incident was a near miss (if applicable).	Free text
Reporter at the scene of the incident: explain in detail what you think caused the incident, including any relationships between contributory factors, include suggestions, comments, and recommendations.	Free text
Manager: explain in detail what you think enabled the incident, including any relationships between contributory factors, include suggestions, comments, and recommendations.	Free text

It should be noted that incident reporters were not required to classify contributory factors; this analysis was undertaken after the incident report was entered into the software tool (described in Section 11.5.1).

In addition to incident data, participation data was collected to enable a calculation of incident rates for different types of LOAs. The participation data included the number of participants and participation days for each LOA conducted during a calendar month.

Participation days represent the number of days on which a participant is exposed to an activity. As long as the activity takes place during a single 24-hour period, then it is counted as a single participation day. This means that any activities undertaken for a single hour (e.g. archery), or 8 or more hours (e.g. bushwalking), were treated equally so long as they are completed in a single day. Similarly, if a participant experienced three LOAs in the same 24-hour period, then each activity would be counted as a single participation day. This is consistent with the calculation of exposure in previous research within residential camps (American Camp Association, 2011), and other outdoor pursuits, such as skiing (Dickson, 2012). While a more precise measure of exposure was included in the prototype UPLOADS (i.e. participation hours), the implementation trial showed that LOA providers were unable to accurately provide this information.

An activity type classification scheme was developed to standardize the collection of participation data across multiple organizations. The scheme was developed to address problems with consistently reporting participation data identified in the implementation trial. The activity type classification scheme, shown in Table 11.5, includes 20 broad activities types.

TABLE 11.5

UPLOADS Activity Types Classification Scheme

Activity Classification	Examples
Archery	–
Arts and crafts	Bush art
Beach activities	Beach sports/activities, fishing, sandboarding
Campcraft	Cooking, campfires
Camping tents	Soft-top accommodations
Caving	Natural, artificial
Curriculum-based activities	Environmental, conservation, science studies
Free time outdoors	Unstructured play
Harness: indoors	Climbing artificial surfaces
Harness: outdoors	Bouldering, giant swing, abseiling, canyoning, flying fox/zip line, prussiking
Horse back/camel back riding	–
Ocean activities	Sailing, sea kayaking, snorkelling, surfing, swimming, stand-up paddle boarding
Residential camps	Hard-top accommodations
River activities	Canoeing, dragon boating, kayaking, rafting, creek dipping
Snow sports	Cross-country/Nordic, downhill, snowboarding
Teambuilding games	Animal games, initiatives/team games, night-time activities
Trampoline	–
Travelling to activity	–
Walking/running outdoors	Adventure racing, bird-watching, bushwalking, geocaching, orienteering, rogaining
Wheel sports	Cycling, mountain biking, quad biking, skating

11.4 The Process for Learning from Incidents

There are two important points of difference between the prototype, and final version of the learning from incidents process shown in Figure 11.2. The first change was to address the finding that activity leaders often did not report details beyond the immediate context of the incident. The new process involved the person supervising, or providing oversight for the entire activity program (i.e. typically a supervisor or field manager), in the incident reporting process. This two-stage data collection approach was intended to ensure that incident reports included any organizational factors influencing the design of the activity program, and external influences, such as regulations, and interactions with schools and parents. It was also anticipated that this would have the added safety benefit of ensuring that those providing oversight for the activity program would be informed of all incidents.

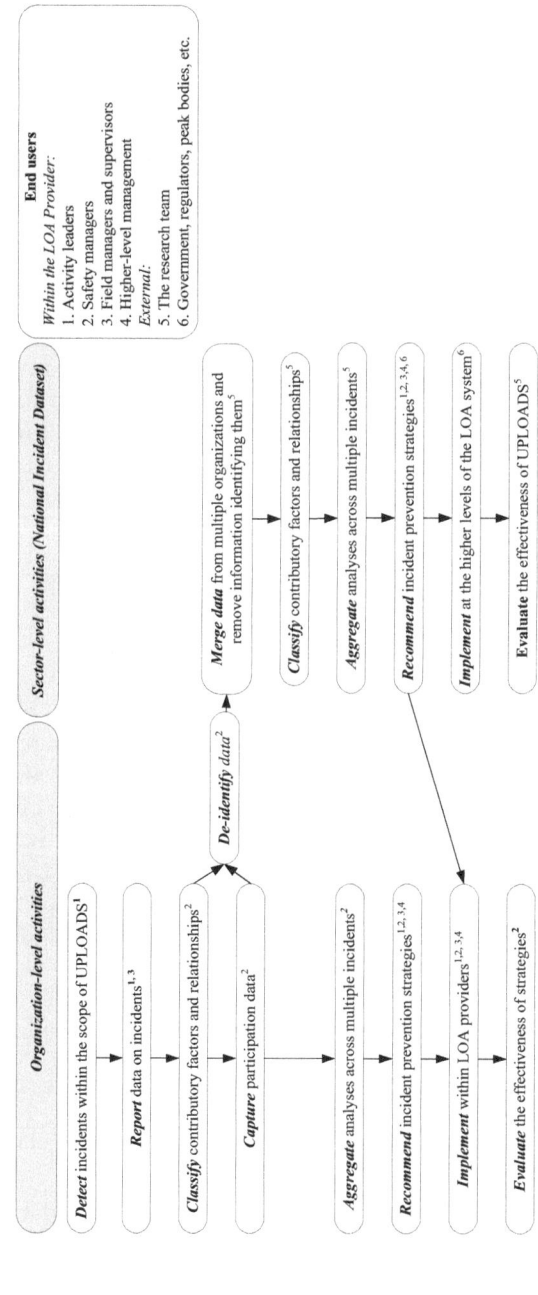

FIGURE 11.2

The learning from incidents processes designed for the UPLOADS.

The second important feature is that all data submitted to the National Incident Dataset (NID) was to be validated and analyzed by the research team, to ensure that the data and analyses are credible, valid and reliable. This process was introduced to address issues relating to the completeness and reliability of the analyses undertaken by safety managers during the implementation trial (described in Chapter 9).

A protocol was also developed to guide the analysis and reporting of results from the NID to ensure privacy and confidentiality for organizations contributing data. The protocol included three core restrictions in reporting the results. First, the details of individual incidents should not be reported in isolation. Second, there must be at least 20 incident reports associated with a particular type of incident to form the basis for an aggregate analysis (e.g. a detailed analysis of only three incidents involving a social/psychological outcome would not be undertaken). Finally, to report on incidents associated with a specific LOA, the activity in question must be delivered by three or more LOA providers. These restrictions were introduced to address the concerns that were raised around confidentiality during the implementation trial.

11.5 Supporting Software Tools

Two software tools were developed: the UPLOADS software tool and UPLOADS Lite. UPLOADS Lite was purely a data collection tool. It was an online survey tool that allowed LOA providers to contribute anonymous incident reports and participation data to the NID, and save the data for their own records. It was developed based on feedback that some LOA providers did not want to analyze the contributory factors associated with their own incident data, and they wanted a simpler system.

In contrast to UPLOADS Lite, the UPLOADS software tool provided a comprehensive approach for collecting, storing, and analyzing data. It provided the following core functions:

1. Collect and store incident report forms and participation data in relational databases;
2. Generate descriptive statistics on the incident data;
3. Analyze incident reports using the contributory factor classification scheme;
4. Produce aggregate Accimaps of the analyzed incident data; and
5. Automatically de-identify data (i.e. names removed) to send to the NID.

The first two functions are common to most incident reporting systems. The unique functions were the analysis of the incident data using the contributory factor classification scheme, and the production of aggregate Accimaps.

11.5.1 The UPLOADS Software Tool Incident Analysis Process

After each incident report form was entered into the Software Tool, the user was prompted to analyze the report. The first section of the interface, shown in Figure 11.3, prompted the identification and classification of the contributory factors in the *Incident Description* fields of the report (see Table 11.4). An unlimited number of factors could be identified.

A hierarchical structure was used to present the classification scheme (shown in Figure 11.1) within the interface. This was implemented to address problems with presenting the classification scheme as a list, identified in the usability evaluation. To classify each contributory factor identified from an incident report, a Level 1 category from the classification scheme was selected from a drop-down box. If possible, a corresponding Level 2 factor was also selected. The Level 2 menu only presented the factors that were relevant to the Level 1 category that had been selected, along with examples of the types of contributory factors that should be categorized under that category.

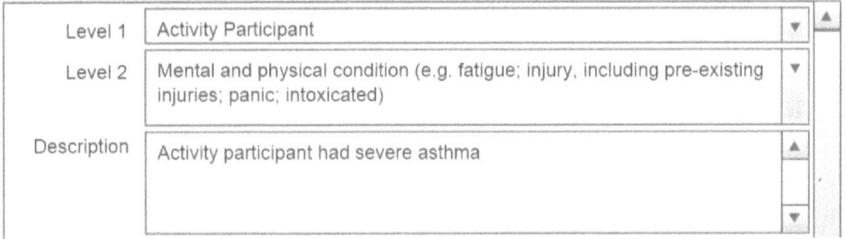

FIGURE 11.3
User interface for identifying and classifying the factors identified from incident reports.

A description of the contributory factor was then entered (e.g. 'Back pack rubbed clients back'). If a description was not entered before moving onto the next factor, the software prompted the user to do so; it was a mandatory requirement. This ensured that all classifications were supported by specific descriptions of the contributory factor identified from the incident report. This was considered important for interpreting the resulting aggregate analyses, and was identified as a limitation of the data collected during the implementation trial (Chapter 10).

A similar procedure was followed for identifying and classifying relationships between contributory factors. This is highly novel for an incident reporting system. The user interface for this process is shown in Figure 11.4. To classify each relationship, a pair of categories was selected from the drop-down menus. Only categories that had been used to classify contributory factors in the report were presented. A description of the relationship was then entered into the 'Description' box. Again, the description was a mandatory requirement in the system.

Once incident reports had been analyzed, the user was able to create Accimap analyses of incident reports. Accimap analyses could be created

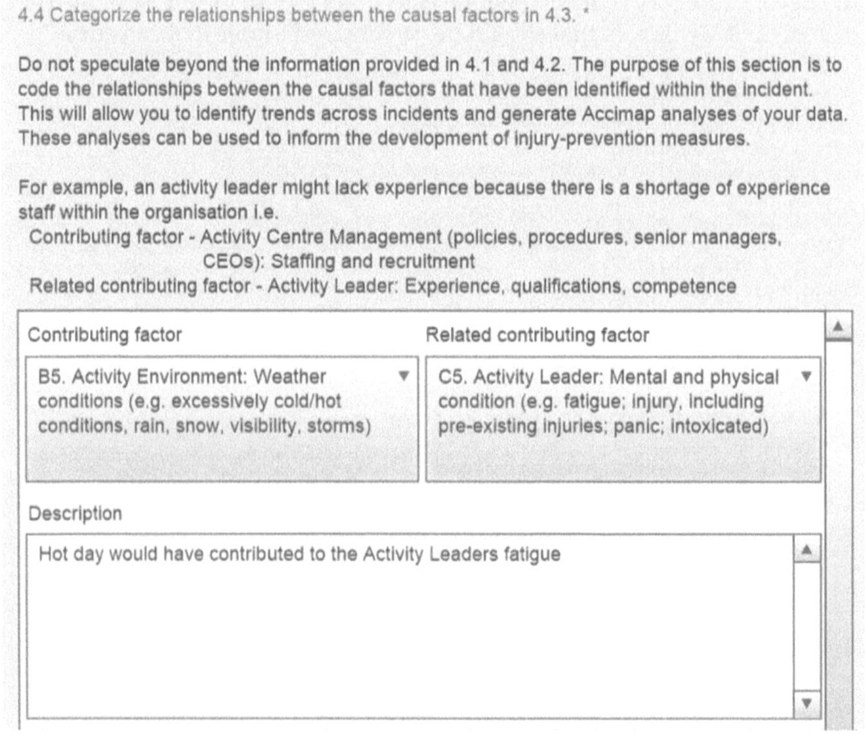

4.4 Categorize the relationships between the causal factors in 4.3. *

Do not speculate beyond the information provided in 4.1 and 4.2. The purpose of this section is to code the relationships between the causal factors that have been identified within the incident. This will allow you to identify trends across incidents and generate Accimap analyses of your data. These analyses can be used to inform the development of injury-prevention measures.

For example, an activity leader might lack experience because there is a shortage of experience staff within the organisation i.e.
    Contributing factor - Activity Centre Management (policies, procedures, senior managers,
                        CEOs): Staffing and recruitment
    Related contributing factor - Activity Leader: Experience, qualifications, competence

Contributing factor	Related contributing factor
B5. Activity Environment: Weather conditions (e.g. excessively cold/hot conditions, rain, snow, visibility, storms) ▼	C5. Activity Leader: Mental and physical condition (e.g. fatigue; injury, including pre-existing injuries; panic; intoxicated) ▼

Description

Hot day would have contributed to the Activity Leaders fatigue

FIGURE 11.4
User interface for identifying and classifying the relationships identified from incident reports.

based on a single incident, all records in the database, or a subset (e.g. only incidents involving a certain activity, or only incidents associated with adverse outcomes). Once the desired incident records had been selected, the software tool produced a graphical summary of the frequency of the contributory factors and relationships identified in the reports. The end user could then select contributory factors within the Accimap (i.e. using the cursor), and the user interface highlighted the identified relationships to other factors.

A text file accompanied the Accimap with a table summarizing all the descriptions of the contributory factors and relationships, and their frequency. An excerpt from a text file of an aggregate analysis of four incident reports is presented in Table 11.6.

11.6 Training Materials

Training materials for UPLOADS were developed for safety managers and incident reporters (i.e. activity leaders and field managers/supervisors).

11.6.1 Safety Manager Training Material

The training materials for safety managers were presented via a website (https://uploadsproject.org/training-material/) with links embedded into the UPLOADS software tool and UPLOADS Lite. The training materials included:

- A manual describing the overall approach to incident reporting and analysis, including the systems thinking approach, and how to collect appropriate incident data;
- A manual describing each component of UPLOADS, including the classification schemes and the software tool;
- A manual describing how to create reports on incident statistics, and interpret the Accimap analyses;
- Online videos demonstrating how to use the software tool (e.g. entering incidents reports, classifying contributory factors and relationships, aggregate analyses).

A series of 1-day workshops using these materials were, conducted across Australia with LOA providers prior to, and during, implementation. Additional workshops on systems thinking were also conducted at regular intervals throughout implementation.

The manual describing the overall approach to incident reporting and analysis is available in Appendix D. This manual could easily be adapted to support an incident reporting system underpinned by systems thinking in any domain.

TABLE 11.6

Excerpt from Text File Produced by the Aggregate Analysis of Four Incident Reports: The Excerpt Shows the Factors Identified at the Local Area Government, Activity Centre Management, Planning, and Budgeting Levels of the Accident Analysis Framework and Two Relationships

Level 3: Local Area Government, Activity Centre Management, Planning and Budgeting

Factor	Occurrences	Descriptions
Higher-Level Management: Policies and procedures for activities and emergencies	4	Incident 1: The appropriate exit was not specified in any activity program documentation. Incident 1: Activity centre documentation did not stipulate limits on group size for activity. Incident 3: The activity centre's policy did not contain information regarding the fact that boots should not be worn for deep water activities. Incident 3: Use of flotation devices in deep water activities were not stipulated in activity centre policy. Incident 4: No consent form or medical information form was used by the activity centre. Incident 4: There was no emergency procedures for the activity in question.
Higher-Level Management: Risk assessment and management	1	Incident 3: There was no risk assessment conducted by the activity centre.
Higher-Level Management: Staffing and recruitment	1	Incident 1: Not enough staff were available to cover the Activity Leader taking a break. Incident 1: New staff had not been recruited to replace staff who had left.

Accimap Relationships

Contributory Factor	Related Contributory Factor	Occurrences	Descriptions
State and Federal Government: Communication	Activity Group Factors: Group size	1	Incident 1: Group size was not limited due to lack of communication from National Parks and Wildlife.
State and Federal Government: Communication	Higher-Level Management: Policies and procedures for activities and emergencies	1	Incident 1: Documentation was not available due to lack of communication from National Parks and Wildlife.

11.6.2 Incident Reporter Training Materials

All the information required to appropriately report an incident was included in the incident report form, and its accompanying training material. Participating organizations were provided with a PowerPoint presentation, and script explaining the purpose of the UPLOADS project, the underpinning theory, what types of incidents to report, and the details required for the incident report forms.

11.7 Summary

This chapter has presented an overview of UPLOADS as of June 2014, and the processes used to ensure that reliable, valid, and credible analyses were produced within organizations, and for the Australian LOA sector. The following chapters describe the findings from the first year of implementing UPLOADS across Australia, and the incident prevention strategies that were identified by the sector based on the findings.

References

American Camp Association. (2011). *Healthy camp study impact report 2006–2010: Promoting health and wellness among youth and staff through a systematic surveillance process in day and resident camps.* Retrieved from http://www.acacamps.org/sites /default/files/images/education/Healthy-Camp-Study-Impact-Report.pdf

Davidson, G. (2005). *Incident Severity Scale. Adapted and expanded from the Accident Frequency Severity Chart (Priest, 1996).* Retrieved from http://www.incident report.org.nz/resources/Severity_Scale.pdf

Dickson, T. J. (2012). Learning from injury surveillance and incident analysis. In T. J. Dickson & T. Gray (Eds.), *Risk management in the outdoors: A whole-of-organization approach for education, sport and recreation* (pp. 204–230). Cambridge University Press: Cambridge, UK.

Goode, N., Salmon, P. M., Taylor, N. Z., Lenné, M. G., & Finch, C. F. (2016). Lost in translation: The validity of a systemic accident analysis method embedded in an incident reporting software tool. *Theoretical Issues in Ergonomics Science, 17*(5–6), 483–506. doi:https://doi.org/10.1080/1463922x.2016.1154230

Leape, L. L., & Abookire, S. (2005). *WHO draft guidelines for adverse event reporting and learning systems: From information to action*: World Health Organization.

Rasmussen, J. (1997). Risk management in a dynamic society: A modelling problem. *Safety Science, 27*(2/3), 183–213.

Svedung, I., & Rasmussen, J. (2002). Graphic representation of accident scenarios: Mapping system structure and the causation of accidents. *Safety Science, 40*(5), 397–s417.

World Health Organization. (1994). *International classification of diseases,* 10th edition (ICD-10). Retrieved from http://apps.who.int/classifications/icd10/browse/2010 /en. Access date: 15th May 2018.

12

Analyzing Incident Data

Practitioner Summary

The power of an incident reporting system underpinned by system thinking lies in the capability to identify the contributory factors, and relationships, involved in multiple incidents. This chapter provides a step-by-step guide to analyzing incident data for this purpose. To illustrate the process, the chapter presents an analysis of the injury data that was collected via UPLOADS by 31 organizations during a one-year period. The findings from the analysis have several important implications for organizations wishing to understand, and prevent, incidents their own incidents.

Practical Challenges

To ensure that all incident data is classified accurately prior to analysis, it is good practice to verify responses to standardized response options (e.g. severity rating scales, whether an injury occurred) against the qualitative information provided in the incident report. This is often a lengthy process, and it reinforces the need to ensure that: (a) standardized response options are comprehensive, and easily interpretable by incident reporters; and (b) objective criteria are used to define the response options.

12.1 Introduction

A well-designed and effectively implemented incident reporting system is a significant resource for safety monitoring and improvement activities. While a single incident report may help to identify new or unexpected hazards, aggregating the data from multiple incident reports is much more powerful.

If the incident reporting system has been designed based on the guidance presented in this book, aggregate data can be used to identify the contributory factors, and relationships between them involved in multiple incidents.

This chapter provides a step-by-step guide to analyzing and interpreting data collected via an incident reporting system underpinned by systems thinking. To illustrate how to report and interpret the findings from such an analysis, an analysis of injury data collected via UPLOADS is presented. The chapter concludes with the broader implications of the findings for organizations wishing to understand and prevent their own incidents.

12.2 Step-by-Step Guide to Analyzing Incident Data

The purpose of this process is to produce an analysis report to inform the development of incident prevention strategies (using the process described in Chapter 13). It is assumed that the incident reports have already been analyzed using an accident analysis method based on Accimap, with an associated contributory factor classification. The process could be adapted to analyze data collected from any incident reporting system, if the contributory factors were suitably categorized.

12.2.1 Define the Scope

The first step is specifying criteria for the types of incident reports to include in the analysis. This could be based on any variable, or combination of variables, in the incident data. At a minimum, the scope should specify a date range (e.g. last quarter, past financial year, past calendar year), and the type of incidents (e.g. injury incidents, illnesses, near misses) to include in the analysis.

The scope needs to be broad enough to ensure confidentiality and privacy for the people involved in incidents. For example, reporting on data from the UPLOADS National Incident Database (NID) was restricted to a minimum of 20 incidents (see Chapter 11). The scope of the analysis needs to be broad enough so that there is not an undue focus on isolated events (and therefore blaming individuals).

The scope also needs to include enough incidents so that it is possible to identify recurring contributory factors; this is highly dependent on the quality of the data that has been collected. If the incident reports include a rich description of the contributory factors involved in the incident, then the analysis may focus on a smaller sample of reports. If the incident reports include a less detailed description (which is typically the case), then recurring themes and patterns may only be evident across many hundreds of reports.

12.2.2 Merge the Data for Analysis

The incident reports, and associated exposure data, then needs to be merged into spreadsheets for the analysis. The incident data spreadsheet should be formatted so that each line in the spreadsheet represents a single incident report. The exposure data spreadsheet should be formatted so that incident rates can be calculated as per Section 12.2.5.

12.2.3 Validate the Data

The next step is assessing the quality of the data, and recoding standardized response options if required. This involves comparing the qualitative description of the incident to responses to standardized response options. At a minimum, this should involve checking that the type of incident (e.g. near miss, adverse outcome, injury, illness, etc.), and severity ratings, have been recorded accurately, because these variables are often used to specify the scope of the analysis.

It is also useful to calculate the percentage of missing responses for each variable, and identify incomplete incident reports. If variables are not reported consistently (i.e. >50% missing data), then these should be excluded from the analysis. At a minimum, incident reports need to include a qualitative description of the incident to be useful for an analysis of the contributory factors.

12.2.4 Filter the Data for the Analysis

Based on the scope of the analysis, filter the spreadsheets, and select the relevant incident reports and exposure data. The subsequent analysis should just focus on the filtered data.

12.2.5 Calculate Incident Rates

To allow comparisons against incidents rates found in other similar settings you should select a method for calculating incident rates based on the industry or context.

For example, in Australia, *incidence rates* and *frequency rates* are recommended for calculating occupational incident rates (Worksafe Australia, 1990). The *incidence rate* is calculated based on the number of reports of injury or disease associated with lost time for each one hundred workers employed. The 'number of workers' excludes workers who were absent for the entire period:

$$\frac{\text{Number of lost time injury/disease cases reported in the recording period}}{\text{Average number of workers in the recording period}} \times 100$$

The *frequency rate* is calculated based on the number of reports of injury or disease associated with lost time for each 1 million hours worked:

$$\frac{\text{Number of lost time injury/disease cases reported in the recording period}}{\text{Number of hours worked in the recording period}} \times 1,000,000$$

Once calculated, these rates can then be compared to the rates for the relevant industry groups in Australia (https://www.safeworkaustralia.gov.au/topics/lost-time-injury-frequency-rates-ltifr). Similar information is available from workplace health and safety regulators internationally.

12.2.6 Describe the Incident Characteristics

Summarize responses to standardized response options using descriptive statistics, and represent the findings using graphs. It is also useful to create cross-tabulation summaries for combinations of variables (e.g. injury location by injury type, or injury type by severity). Organize the findings using the following headings to provide an overview of the characteristics of the incidents included in the analysis:

- *Contextual factors* at the time of the incidents;
- *Demographic information* on the people impacted by the incidents; and
- *Outcomes* from the incident.

As with the incident rates, the findings should be presented with an interpretation of their meaning. This might involve comparing the findings to a previous reporting period or research findings, or highlighting potential problems.

12.2.7 Construct Summary Accimaps with Examples

The next step involves representing the types of contributory factors and relationship identified from multiple incident reports as an Accimap (e.g. a diagram) using drawing software, such as Microsoft Visio. The 'types' should be based on the categories in the contributory factors classification scheme. Figure 12.1 shows a template for an Accimap, summarising the number and percentage of incident reports identifying each contributory factor.

To enhance readability, it is useful to construct two Accimaps summarizing:

1. The frequency of the types of contributory factors identified in the incident reports (see Figure 12.4); and
2. The frequency of the types of relationships identified in the incident reports (see Figure 12.5).

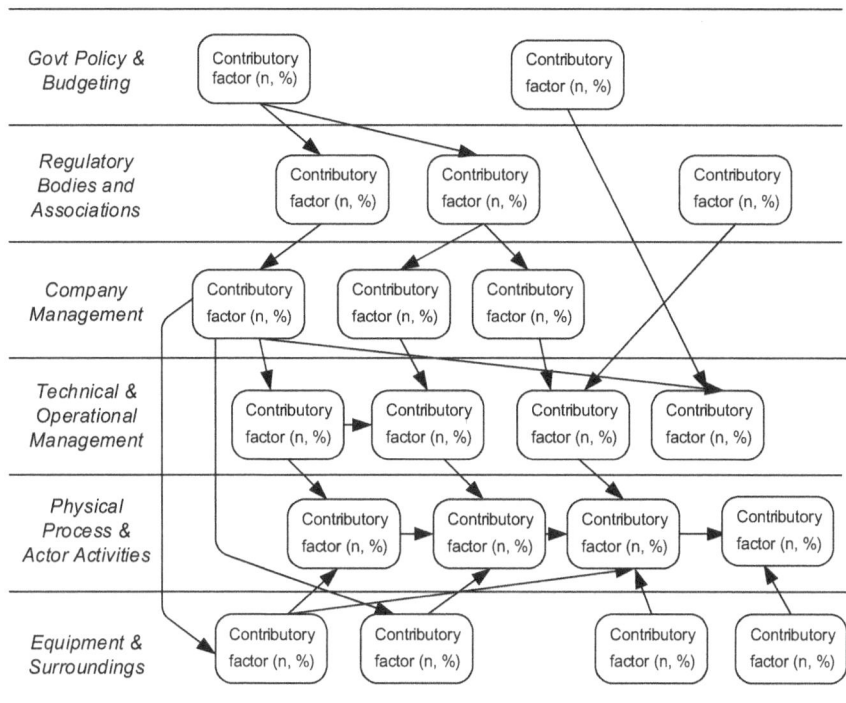

FIGURE 12.1

Example of using the Accimap framework to summarize the frequency (n) and percent (%) of incident reports identifying specific types of contributory factors.

To appropriately interpret the Accimaps, they need to be accompanied by tables providing quotes from incident reports illustrating the types of contributory factors and relationships (see Tables 12.2 and 12.3).

12.2.8 Describe and Interpret the Accimap

Describe the findings from the Accimaps and tables in text. Start by describing the findings at the top level of the Accimap framework (e.g. government policy and budgeting), and work downwards. For each level of the Accimap framework, describe: (1) the most frequently identified types of contributory factors; (2) the relationships identified within and across levels; and (3) the implications of the findings for risk management.

12.2.9 Use the Analysis to Develop Incident Prevention Strategies

Finally, and most importantly, the findings from the analysis should then be used to develop incident prevention strategies. Chapter 13 provides detailed guidance on translating incident analyses into strategies that align with systems thinking.

12.3 Analysis of Data Collected via UPLOADS

This section describes the findings from the injury data collected via UPLOADS during the first 12 months of implementation (June 2014 to May 2015). It is intended to provide readers with an example of the outputs from the process described in Section 12.2. The development of incident prevention strategies based on this analysis is described in Chapter 13.

12.3.1 Scope

The analysis included data from 31 organizations. The scope of the analysis was limited to injuries reported from 1st June 2014 to 31st May 2015.

12.3.2 Data Merging Validation and Filtering

The incident and participation data from each organization were merged into two spreadsheets.

The actual severity ratings, incident type (near miss or adverse outcome), and adverse outcomes (injury, illness, equipment damage) recorded in each incident report were validated against the incident description, and recoded if required. The incident reports associated with injuries were selected, and the final dataset included 676 incident reports.

An initial check of the participation data found that some values for participation days were outside the range of possible values, indicating that data had been entered incorrectly. The total number of participants per activity was therefore used as a denominator for incident rate calculations (i.e. rate per 1,000 participants).

12.3.3 Incident Rates

Annual incident rates were calculated per 1,000 participants for each activity using the formula:

$$\frac{\text{Number of incident reports associated with each activity}}{\text{Number of people participating in the activity}} \times 1,000$$

The rates were then averaged across all activities to provide an estimate of the overall rate.

Across all activities, the annual incident rate was 2.1 per 1,000 participants. This rate is broadly consistent with findings from international studies of LOA injury, which report incident rates ranging from 0.54 to 2.30 per 1,000 participation days (Barst, Bialeschki, & Comstock, 2008; Leemon & Schimelpfenig, 2003). While the findings from these international studies are not directly comparable with our Australian data due to the format of reporting participation data (i.e. per 1,000 participation days as opposed to per 1,000 participants),

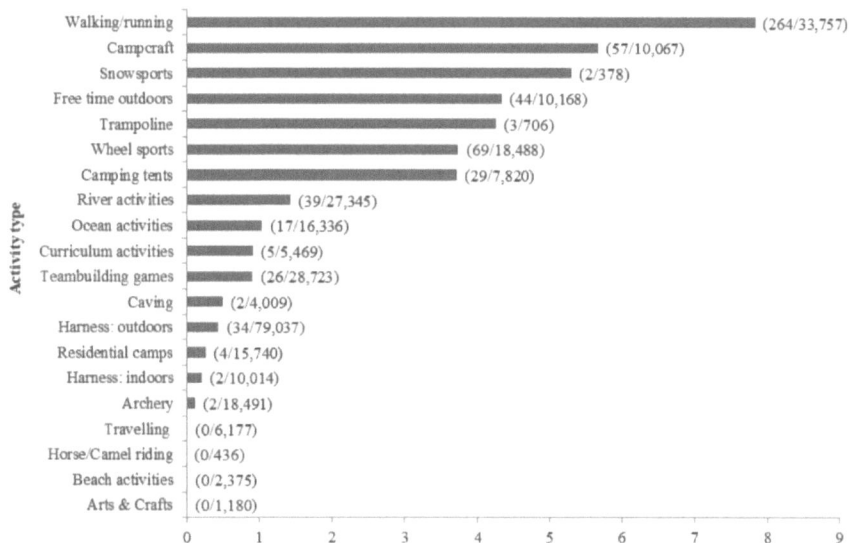

FIGURE 12.2
Incident rates per 1,000 participants by activity type. Numbers in brackets represent the number of incident reports and the number of participants recorded for the activity.

overall the findings support the conclusion that the relative frequency of reported injury in Australia is similar to international contexts.

As shown in Figure 12.2, walking and running in the outdoors had the highest reported incident rate, followed by campcraft (i.e. cooking, campfires) and snow sports. This suggests that the risk management strategies in place to reduce injuries associated with walking and running activities (e.g. bushwalking) need to be reviewed by organizations providing these activities. Similarly, campcraft and free time should also be reviewed, as they were also associated with a relatively high incident rate, and large numbers of participants.

More than half of all activity types had an incident rates of less than 1 per 1,000 participants, including many typically perceived as 'high risk' by people unfamiliar with outdoor activities (e.g. caving, indoor and outdoor harness activities, kayaking). From a risk management perspective, this suggests that these activities are well controlled by the organizations contributing data. A wider implication is that risk management practices may be overly focused on ostensibly risky activities, and focused less so on activities that are perceived to be low risk, or that are not treated as part of the LOA program (e.g. walking, campcraft, free time). This is indicated by the fact that the latter type of activities had the highest incident rates.

12.3.4 Incident Characteristics

Descriptive statistics for incident characteristic variables were calculated using a statistical software package (SPSS) (version 21).

12.3.4.1 Contextual Factors

The majority of injuries occurred when there was an absence of rain (75%, n = 507); a moderate temperature (71%, n = 479), and no or minimal wind (87%, n = 588).

There was a ratio of 1 activity leader for every 8 participants in activities associated with injuries. The activity leader was reported to have relevant qualifications for the activity in the majority of incidents (80%, n = 540).

12.3.4.2 Demographic Information

Most injured activity participants were female (227 female, 91 male), with a median age of 16 years (range 10 to 18 years). Demographic information was not collected with the participation data, so it is unknown whether this difference is attributable to initial gender differences in participation.

Gender was relatively evenly split for injured people within the roles of activity leader (21 female, 20 male), field/program managers (10 female, 3 male), teachers (4 female, 4 male), and volunteers and interns (9 female, 11 male). Injured persons in these supervisory roles had a median age of 26 years (range 15 to 60).

12.3.4.3 Outcomes

A summary of the type and frequency of injuries sustained according to body location is shown in Figure 12.3. Ankles and feet were most frequently injured (35%, n = 233). While most of these injuries were classified as superficial, there was also a relatively high number of dislocations, sprains, and strains. Another significant trend was the relatively high rate of burns and corrosions across body locations. This reinforces the need for organizations to review risk management strategies for campcraft (e.g. cooking and campfire) activities.

Most reported injuries (85%, n = 577) were minor, requiring only first aid and no evacuation. The frequency of injuries sustained according to body location and severity is shown in Table 12.1. The most severe injuries were to the head; these were mainly burns and corrosions, and eye injuries.

12.3.5 Accimap Analysis

Out of the 676 incident reports involving injuries, 363 had sufficient details regarding contributory factors. On average, only 2.2 contributory factors were identified per incident report. Relationships between factors were only identified in 55 reports, with an average of one per report. The remaining reports often only included details about the actual injury that occurred. This indicates that further education is needed around the level of detail required in incident reports to support the analysis.

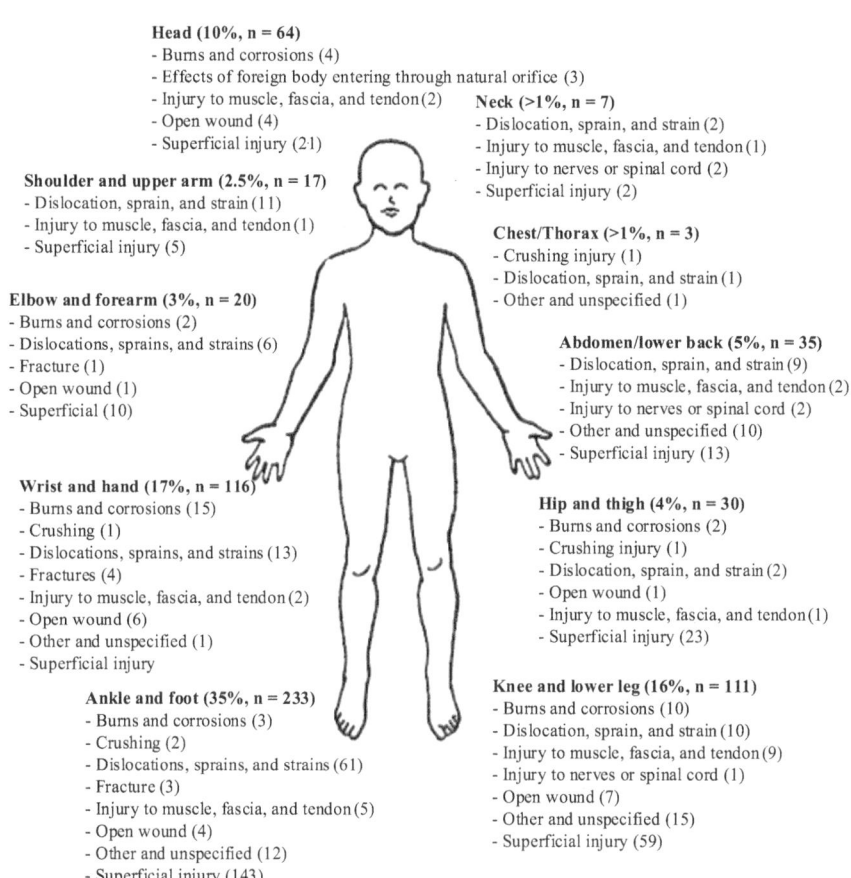

Head (10%, n = 64)
- Burns and corrosions (4)
- Effects of foreign body entering through natural orifice (3)
- Injury to muscle, fascia, and tendon (2)
- Open wound (4)
- Superficial injury (21)

Neck (>1%, n = 7)
- Dislocation, sprain, and strain (2)
- Injury to muscle, fascia, and tendon (1)
- Injury to nerves or spinal cord (2)
- Superficial injury (2)

Shoulder and upper arm (2.5%, n = 17)
- Dislocation, sprain, and strain (11)
- Injury to muscle, fascia, and tendon (1)
- Superficial injury (5)

Chest/Thorax (>1%, n = 3)
- Crushing injury (1)
- Dislocation, sprain, and strain (1)
- Other and unspecified (1)

Elbow and forearm (3%, n = 20)
- Burns and corrosions (2)
- Dislocations, sprains, and strains (6)
- Fracture (1)
- Open wound (1)
- Superficial (10)

Abdomen/lower back (5%, n = 35)
- Dislocation, sprain, and strain (9)
- Injury to muscle, fascia, and tendon (2)
- Injury to nerves or spinal cord (2)
- Other and unspecified (10)
- Superficial injury (13)

Wrist and hand (17%, n = 116)
- Burns and corrosions (15)
- Crushing (1)
- Dislocations, sprains, and strains (13)
- Fractures (4)
- Injury to muscle, fascia, and tendon (2)
- Open wound (6)
- Other and unspecified (1)
- Superficial injury

Hip and thigh (4%, n = 30)
- Burns and corrosions (2)
- Crushing injury (1)
- Dislocation, sprain, and strain (2)
- Open wound (1)
- Injury to muscle, fascia, and tendon (1)
- Superficial injury (23)

Ankle and foot (35%, n = 233)
- Burns and corrosions (3)
- Crushing (2)
- Dislocations, sprains, and strains (61)
- Fracture (3)
- Injury to muscle, fascia, and tendon (5)
- Open wound (4)
- Other and unspecified (12)
- Superficial injury (143)

Knee and lower leg (16%, n = 111)
- Burns and corrosions (10)
- Dislocation, sprain, and strain (10)
- Injury to muscle, fascia, and tendon (9)
- Injury to nerves or spinal cord (1)
- Open wound (7)
- Other and unspecified (15)
- Superficial injury (59)

FIGURE 12.3

Type and frequency of injuries sustained according to body location. (From van Mulken, M., Clacy, A., Grant, E., Goode, N., Finch, C. F., Stevens, E., & Salmon, P. M. 2016. *The UPLOADS National Incident Dataset the first twelve months: 1st June 2014 to 31st May 2015.*)

The frequency and percent of the types of contributory factors identified across the incident reports was calculated, and represented on the adapted Accimap framework (shown in Figure 12.4). A separate Accimap was produced representing the frequency and percent of the types of relationships identified across the incident reports (shown in Figure 12.5). To support the interpretation of the Accimaps, tables were produced using quotes from the incident reports to illustrate the types of contributory factors, and relationships, identified.

As Figures 12.4 and 12.5 show, no factors were identified at the upper two levels of the Accimap framework (i.e. Government departments; Regulatory bodies and associations).

The most frequently identified factors at the **Local Area Government, Schools, Parents and Carers, and Higher-Level Management level** related

TABLE 12.1

Severity Rating and Frequency of Injuries Sustained According to Injury Location

Injury location	Minor, No Evacuation		Moderate, External Care or Evacuation		Serious+, Timely Evacuation, Emergency Services	
	%	N	%	N	%	N
Head	8	47	13	10	8	6
Neck	1	5	1	1	1	1
Shoulder, Upper arm	2	14	3	2	3	2
Chest, Thorax	<1	2	1	1	–	–
Abdomen, Lower back, Lumbar spine, Pelvis	4	25	7	5	3	2
Wrist, Hand	18	103	14	11	4	3
Elbow, Forearm	2	13	4	3	3	2
Hip/Thigh	5	30	3	2		
Knee/Lower leg	16	93	24	18	1	1
Ankle/Foot	37	213	22	17	4	3
Multiple body regions	2	11	1	1	3	2
Unspecified	4	21	7	5	–	0
Total		577		76		22

Source: From van Mulken, M., Clacy, A., Grant, E., Goode, N., Finch, C. F., Stevens, E., & Salmon, P. M. 2016. *The UPLOADS National Incident Dataset the first twelve months: 1st June 2014 to 31st May 2015.* With permission.

Note: Boxes shaded in grey indicate most commonly injured body location in each severity rating category.

to the management of the organization, including training and evaluation of staff (2.5%, n = 9); policies and procedures for activities and emergencies (1.7%, n = 6); and risk assessment and management (1.4%, n = 5). Specific examples of these contributory factors from the incident reports are shown in Table 12.2. These factors highlight the importance of formal systems of management and risk assessment for organizations providing LOAs. The factors relating to the training of staff suggest that organizations may need to examine whether there are currently effective processes in place for: (a) identifying training needs and designing training for LOA programs; and (b) evaluating staff competencies and knowledge.

The reports identified several relationships between training and evaluation of staff and other factors (shown in Table 12.3). These relationships illustrate the flow on effects of policies and procedures onto training, and subsequent activity leader decision making and behavior during LOAs.

The most frequently identified factors at the **Supervisory and Management Decisions and Actions level** were: Activity or program design (11%, n = 40); supervision of activity leaders and other staff (<1%, n = 2); supervision/oversight of programs/activities (<1%, n = 3).

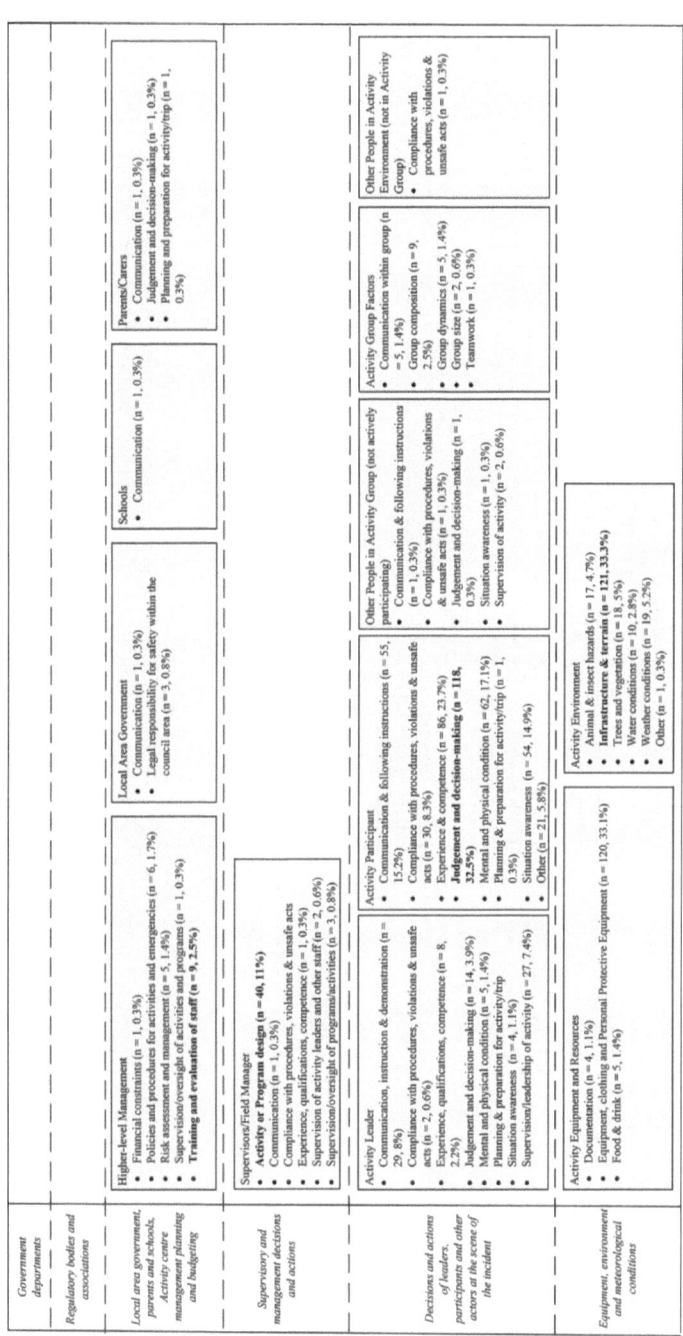

FIGURE 12.4

Frequency of factors identified as contributory to injuries (n = 363 incidents). The most frequently identified factors at each level are highlighted in bold.

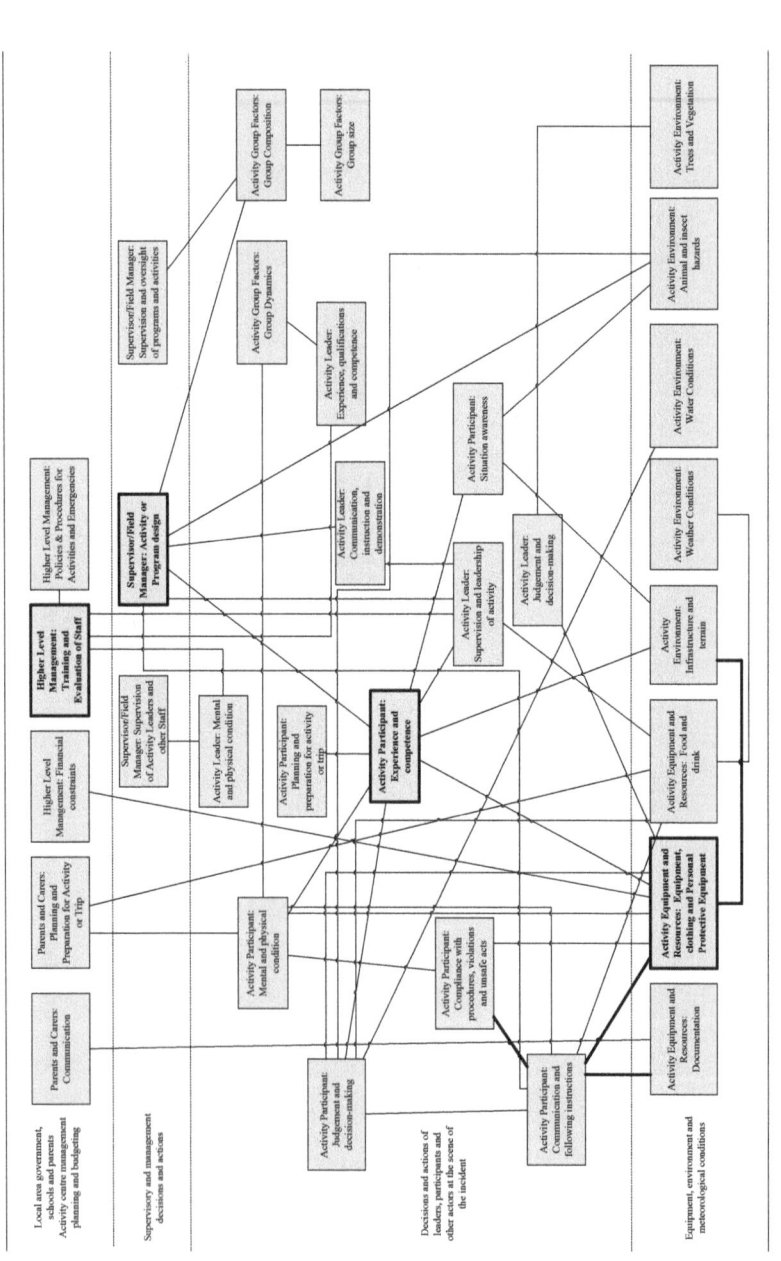

FIGURE 12.5
Frequency of relationships between factors identified as contributory to injuries (n = 55). Interactions identified in more than one report are highlighted in bold. The factors at each level with the most frequent links to other factors are highlighted.

TABLE 12.2

Examples of Contributory Factors Relating to the Management of the LOA Provider

Contributory Factor	Examples from the Incident Reports
Training and evaluation of staff	'Activity leader was not trained in assessing conditions during low tide'
	'There is possibly a need to do further training on how to properly lift canoes safely'
	'Higher management needs to stress supervision responsibilities to experienced staff surrounding working with practical students'
Policies and procedures for activities and emergencies	'Review the standard operating procedures and implement gloves on any hot weather day or closure of the tower for climbing'
	'No communication system in place in the event of miss fire/ incorrect use or emergency situation'
	'There is no standard operating procedure for driver training'
Risk assessment and management	'Risk assessment did not recognize the dangers of a low tide'
	'Neither management or instructor identified the risk . . . even though they were aware of the change in cable tension'

TABLE 12.3

Examples of Relationships between Training and Evaluation of Staff and Other Factors

Relationship		Example Description from Incident Reports
Policies and procedures for activities and emergencies	Training and evaluation of staff	Policies and procedures do not include requirement for manual handling training, leading to a lack of training for staff in this area
Training and evaluation of staff	Activity leader: Supervision and leadership of activity	Training does not clearly state responsibilities for supervising practical students in the field, leading to poor supervision
Training and evaluation of staff	Activity leader: Qualifications and competence	Leader not trained in the assessment of water conditions, leading to an inadequate assessment of water conditions
Training and evaluation of staff	Activity leader: Mental and physical condition	Manual handling training for staff not inclusive of variable physical capabilities (e.g. height), leading to inappropriate lifting practices

The contributory factors relating to activity or program design reflect issues relating to the scheduling of activities and poor tailoring of programs to participants (shown in Table 12.4). The latter issue is also reflected in the relationships identified between program design factors and participant characteristics (shown in Table 12.5). This finding reflects a similar issue identified in analyses of fatal incidents whereby participants are undertaking activities that are not suited to their capabilities (Salmon, Cornelissen, & Trotter, 2012; Salmon, Williamson, Lenné, Mitsopoulos-Rubens, & Rudin-Brown, 2010). These findings suggest that there is a need for organizations review the processes used to match the design of LOA programs to the specific capabilities of participants.

The most frequently identified factors at **the Decisions and Actions of Leaders, Participants, and Other Actors at the Scene of the Incident level** related to the activity participant judgement and decision-making (32.5%, n = 118); experience and competence (23.7%, n = 86); and mental and physical condition (17.1%, n = 62). Specific examples of these contributory factors from the incident reports are shown in Table 12.6.

The examples relating to judgement and decision making and experience/competence suggest that injuries to participants are commonly associated with unfamiliar activities. This finding, in combination with the factors relating to activity and program design, suggests that the risk of injury may be reduced through the appropriate sequencing of activities during programs, so that participants develop skills or competencies in the activities gradually.

TABLE 12.4

Examples of Contributory Factors Relating to Activity or Program Design

Contributory Factor	Examples from the Incident Reports
Fatigue and tiredness due to length of program or activities	'Group had been walking for over 8 hours' 'Last day of program so participant was feeling tired'
Preparation of participants for later stages of the activity	'Students had only had 1-hour skills session and on water mentoring' 'Lacked the consolidation provided from the normal program sequence from cooking on fires on repeated occasions' 'Better warm up may have lessened the injury'
Activity inappropriate for participant age or ability	'Activity needed to be designed better for age group' 'Need to adapt instructions and carrying conditions as per groups age'
Timing of activity	'Natural light fading' 'Orienteering not a great option as first session due to understanding of the group'
Game increased danger of injury	'Doing blindfold walk' 'Strenuous climb'
Environmental factors impacting on suitability of activity (e.g. slippery, cold)	'Choice of activity not suitable under slippery conditions'

TABLE 12.5

Examples of Relationships between Activity or Program Design and Other Factors

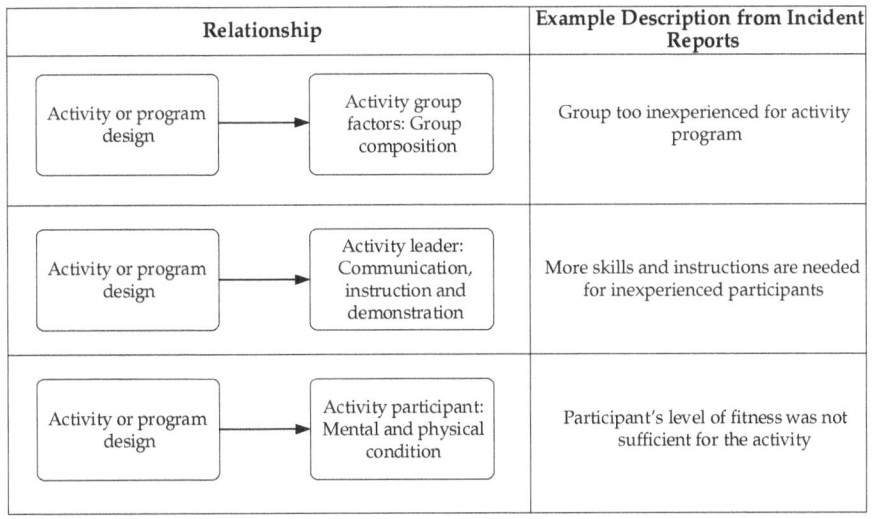

Relationship		Example Description from Incident Reports
Activity or program design	Activity group factors: Group composition	Group too inexperienced for activity program
Activity or program design	Activity leader: Communication, instruction and demonstration	More skills and instructions are needed for inexperienced participants
Activity or program design	Activity participant: Mental and physical condition	Participant's level of fitness was not sufficient for the activity

TABLE 12.6

Examples of Contributory Factors Relating to the Activity Participants

Contributory Factor	Examples from the Incident Reports
Judgement and decision making	'Cycling too close to person in front' 'Lack of judgement on heat and hazard of fires'
Experience and competence	'Exceeded ability and technique' 'Participant perhaps inexperienced at cross country and running whilst map reading'
Mental and physical condition	'The elderly man had been struggling at the beginning of the walk' 'Participant tripped and aggravated pre-existing knee injury'

In addition, the examples relating to the mental and physical conditions of the participants further emphasizes the importance of matching the participant to the program, during the program planning and design phase.

Contributory factors relating to activity leaders were also frequently identified, including: supervision and leadership of activity (7.4%, n = 27); communication, instruction, and demonstration (8%, n = 29); and judgement and decision making (n = 14, 3.9%) (see Table 12.7).

Several relationships were also identified between these factors and other factors influencing the conduct of the activity (see Table 12.8). These relationships illustrate the critical role that activity leaders play in ensuring that the risks associated with providing activities to inexperienced activity participants are appropriately monitored and managed.

TABLE 12.7

Examples of Contributory Factors Relating to the Activity Leaders

Contributory Factor	Examples from the Incident Reports
Supervision/ leadership of activity	'Instructors not being diligent with their supervision' 'Lack of supervision of assistant instructor' 'Unsupervised (broken into smaller groups across site)'
Communication, instruction, and demonstration	'The briefing could have been more involved. Key areas such as maintaining space between riders could have been emphasized'. 'Field staff needed to tap in more often and make sure everything was ok'
Judgement and decision making	'The instructors did not think to stop the walk based on the frailty of the elderly participants and the amount of debris on the path' 'Participant was allowed to continue with heavier load, despite advice'

TABLE 12.8

Examples of Relationships between the Activity Leader and Other Factors

Relationship		Example Description from Incident Reports
Activity Leader: Supervision and Leadership of Activity	→ Activity Leader: Judgement and Decision making	Participant not removed from group because of lack of supervision
Activity Leader: Supervision and Leadership of Activity	→ Activity Participant: Experience and Competence	Participant leading group was walking too fast for others
Activity Leader: Supervision and Leadership of Activity	→ Activity Leader: Communication, Instruction, and Demonstration	Poor communication leading to poor supervision around fires
Activity Leader: Communication, Instruction, and Demonstration	→ Activity Leader: Experience, Qualifications, and Competence	Instructor not experienced to stop activity and provide more instruction to participants
Activity Leader: Communication, Judgement, and Decision making	→ Activity Equipment and Resources: Equipment, Clothing, and Personal Protective Equipment	Forgot to hand the retrieval rope to participants

TABLE 12.9

Examples of Contributory Factors Relating to the Equipment, Environment, and Meteorological Conditions

Contributory Factor	Examples from Actual Incident Reports
Infrastructure and terrain	'Slippery abseil in a canyon that hadn't seen much traffic during the winter months and was quite mossy'
	'The area where tubing is run had seen severe sand erosion leading to the concrete footing of the wall being exposed'
Equipment, clothing, and personal protective equipment	'Trip may have been exacerbated by heavy load'
	'After inspection, found that the gas pump was not connected properly'
Weather conditions	'The weather was very hot'
	'At 5 am it became extremely windy'

Finally, the most frequently identified factors at the **Equipment, Environment, and Meteorological Conditions level** were: infrastructure and terrain (33.3%, n = 121); equipment, clothing, and personal protective equipment (33.1%, n = 120); and weather conditions (5.2%, n = 19) (see Table 12.9). These factors illustrate the importance of reviewing the design of the program and the adequacy of risk control measures directly prior to the activity, as the weather and terrain can change quickly.

12.4 Implications for Understanding and Preventing Incidents

The previous section presented an analysis of the injury data collected via UPLOADS during a 12-month period by 31 organizations in the LOA sector. This section considers the implications of the findings for any organization wishing to understand and prevent incidents.

Perhaps most importantly, the analysis clearly illustrates that collecting data on relatively minor injuries provides a basis for understanding the complex network of contributory factors involved in incident causation. Despite the seemingly benign nature of the injuries reported via UPLOADS, it is evident that they share common systemic causes with other fatal incidents in the LOA domain. For example, the analysis of the Mangetepopo Gorge tragedy presented in Chapter 2 highlighted the role of a poorly designed program, underestimation of the risks associated with the program, a poor risk management system, inadequate and unclear procedures, inadequate training systems, a failure to cancel the trip in response to weather conditions, missing documentation, and activity leader inexperience. These same factors were identified in the reports of minor injuries collected via UPLOADS.

This supports the view that all incident reports, regardless of severity, are a valuable source of information for organizations. Organizations should focus on building a culture where the reporting and in-depth analysis of minor incidents and near misses is actively encouraged and rewarded. The information gleaned through these reports may provide clues which help prevent fatal or catastrophic events.

Another important finding is that the contributing factors involved in incidents do not appear to be activity specific. This suggests that the most effective approach to preventing incidents is to address problems with the overall system of work, rather than focusing on fixing problems associated with specific tasks, work processes, or activities. For example, the analysis identified multiple incidents associated with staff training (e.g. representing problems with assessing conditions during low tide, supervising practical students and lifting canoes). Examined in isolation, the analysis of these incidents would likely result in retraining individual staff, emails to staff about their roles and responsibilities, or introducing new training for specific work activities. A more holistic view of these incidents suggests that the processes used to identify training needs, and design training for activity leaders, need to be examined. If these processes are not redesigned, then it is likely that future incidents will identify additional training gaps for activity leaders, resulting in the need for more training.

More generally, the implication is that aggregating and analyzing incident data across multiple work activities will provide insight into systemic problems. While addressing these problems is often costly and time consuming, it will ultimately result in greater, more long-term reductions in incidents compared with addressing each incident in isolation.

Finally, the findings support the view that sector-wide incident reporting systems are a valuable source of information for proactive risk management. While the number of reports collected by individual organizations was relatively small, this resulted in a reasonably large sample of detailed incident reports. This supported an analysis that was far more detailed than would have been possible from an analysis of data from a single organization. This allows individual organizations to identify potential problems that have not yet occurred in their organization. The implication for other industries and domains is that committing to the development and implementation of a sector-wide incident reporting system underpinned by systems thinking is well worth the effort, due to the potential to prevent, or reduce the impact of, incidents before they occur.

12.5 Summary and Next Steps

This chapter described a step-by-step process for analyzing the findings from an incident reporting system underpinned by systems thinking. This is intended to provide a guide for organizations wishing to analyze their own data. The example analysis clearly illustrates the insights that can be achieved through the sector-wide collection and analysis of incident reports. More generally, the findings suggest that incident prevention strategies should focus on the overall system of work, rather than on specific tasks, work process, or activities. The following chapter provides step-by-step guidance on using incident analyses to design incident prevention strategies.

Further Reading

Examples of aggregate systems analyses in other domains can be found in these papers:

- Newnam, S., & Goode, N. (2015). Do not blame the driver: A systems analysis of the causes of road freight crashes. *Accident Analysis and Prevention*, 76, 141–151.
- Leveson, N., Samost, A., Dekker, S., Finkelstein, S., & Raman, J. (2016). A systems approach to analyzing and preventing hospital adverse events. *Journal of Patient Safety*. doi:10.1097/PTS.0000000000000263.

References

Barst, B., Bialeschki, M. D., & Comstock, R. D. (2008). Healthy camps: Initial lessons on illnesses and injuries from a longitudinal study. *Journal of Experiential Education*, 30(3), 267–270.

Goode, N., Salmon, P. M., Taylor, N. Z., Lenné, M. G., & Finch, C. F. (2016). Lost in translation: The validity of a systemic accident analysis method embedded in an incident reporting software tool. *Theoretical Issues in Ergonomics Science*, 17(5–6), 483–506. doi:https://doi.org/10.1080/1463922x.2016.1154230

Leemon, D., & Schimelpfenig, T. (2003). Wilderness injury, illness, and evacuation: National Outdoor Leadership School's incident profiles, 1999–2002. *Wilderness & Environmental Medicine*, 14(3), 174–182.

Salmon, P. M., Cornelissen, M., & Trotter, M. J. (2012). Systems-based accident analysis methods: A comparison of Accimap, HFACS, and STAMP. *Safety Science*, 50(4), 1158–1170.

Salmon, P. M., Williamson, A., Lenné, M., Mitsopoulos-Rubens, E., & Rudin-Brown, C. M. (2010). Systems-based accident analysis in the led outdoor activity domain: Application and evaluation of a risk management framework. *Ergonomics*, 53(8), 927–939. doi:10.1080/00140139.2010.489966

van Mulken, M., Clacy, A., Grant, E., Goode, N., Finch, C. F., Stevens, E., & Salmon, P. M. (2016). *The UPLOADS National Incident Dataset the first twelve months: 1st June 2014 to 31st May 2015*. Retrieved from https://uploadsproject.files.word press.com/2016/03/first-12-month-report_final.pdf. Access date 15th May 2018.

Worksafe Australia. (1990). Workplace injury and disease recording standard. *Worksafe Australia*.

13

Designing Incident Prevention Strategies

Practitioner Summary

Systems thinking is based on the view that incidents are caused by multiple, interacting, contributory factors. To understand and prevent incidents, organizations need to ensure that this view is embedded in the way they analyze incident data, and in the way they design incident prevention strategies. This chapter provides a step-by-step guide to designing incident prevention strategies using a systems thinking–based approach. The process involves stakeholders from across the levels of a system working together to identify incident prevention strategies. The application of the process is illustrated by describing the incident prevention strategies that were developed by stakeholders based on an analysis of injury data collected via UPLOADS.

Practical Challenges

It is challenging to involve stakeholders from all levels of a system in the design of incident prevention strategies, especially those at the higher levels (e.g. regulators and government agencies). Readers wishing to design incident prevention strategies for their own organization should focus on ensuring that senior managers and executives are involved in the process, as well as line managers and front line workers.

13.1 Introduction

The systems thinking approach suggests that organizations need to address incidents holistically, rather than just attempting to 'fix' broken parts of the system (Dekker, 2011; Hollnagel, 2012; Leveson, 2011). Using Accimap to analyze incident reports is a good starting point, as it provides an understanding of the multiple, interacting, factors involved in incidents across the levels of a system (as shown in Chapter 12). It is often difficult, however, to identify and translate Accimap analyses directly into incident prevention strategies, due to their complexity.

One obvious solution is to focus on the most frequently identified contributory factors in the Accimap. The problem with this solution is that the most frequently identified factors are always related to staff behavior, equipment, and the environment, as these are common to all incidents. Contributory factors at the higher levels of the system are typically identified far less frequently. This approach to designing incident prevention strategies will therefore have a minimal impact on improving safety in the long term.

This chapter provides a step-by-step guide to designing incident prevention strategies using a systems thinking–based approach. The chapter describes a participatory design process for translating Accimap analyses of incidents into strategies that align with Rasmussen's (1997) risk management framework. The process uses the Accimap method to graphically represent the relationships between incident prevention strategies; the outputs are referred to as Preventimaps. The application of the process is illustrated by describing the incident prevention strategies that were developed by stakeholders based on an analysis of injury data collected via UPLOADS.

13.2 Step-by-Step Guide to Designing Incident Prevention Strategies

This process is based on the guidance provided in a systems thinking–based design toolkit (Read, Beanland, Lenné, Stanton, & Salmon, 2017), which was originally developed for use with the Cognitive Work Analysis (CWA) framework (Vicente, 1999). An overview of the design process is shown in Figure 13.1.

13.2.1 Step 1: Define the Desired Outcome

The first step is defining the desired outcome in consultation with key stakeholders (e.g. managers or regulators). The desired outcome might be very broad (e.g. reduce injuries), or very specific (e.g. address issues associated with a particular work activity).

The key is to ensure that everyone agrees on the desired outcome, so that there have clear expectations about the purpose of the process. The desired outcome can also be used to measure the success of the process over the long term.

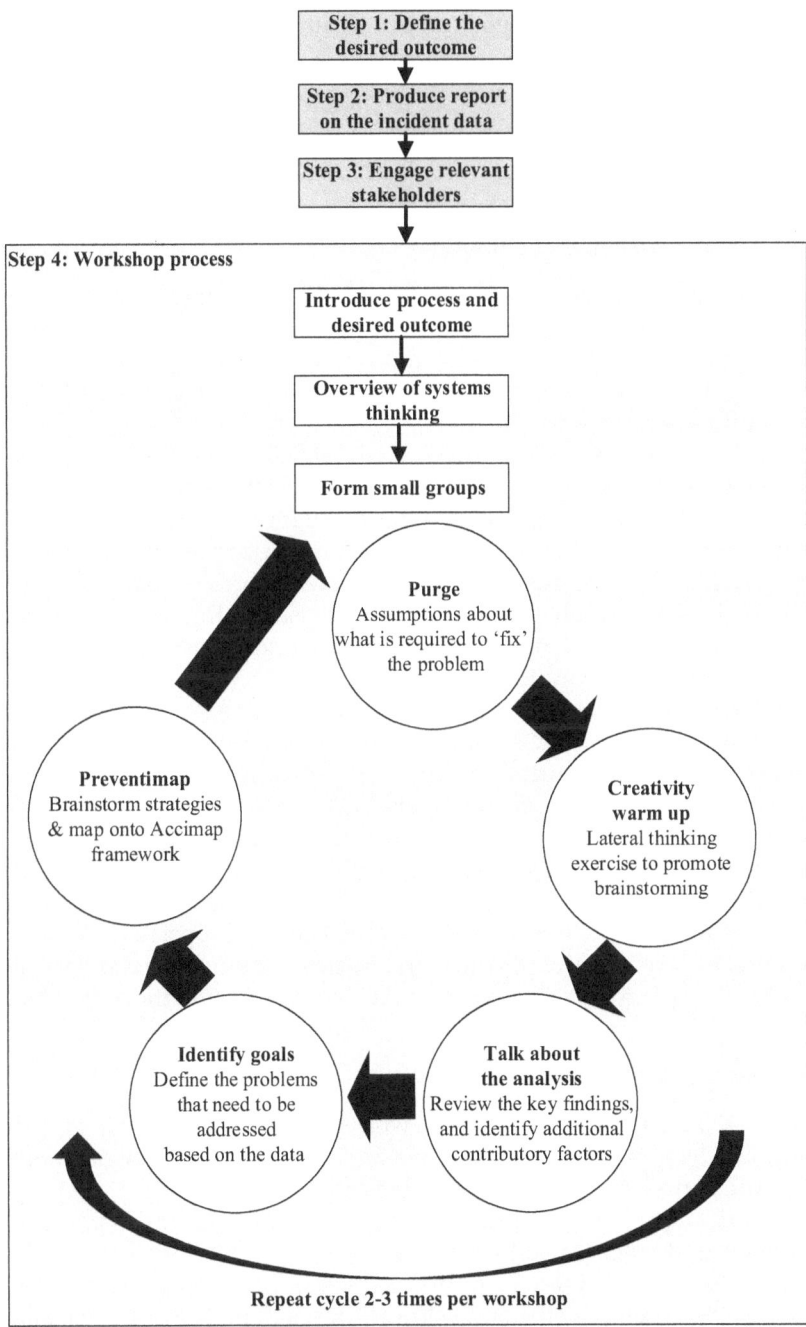

FIGURE 13.1
Incident prevention strategy workshop design process.

13.2.2 Step 2: Produce a Report on the Incident Data

The second step is analyzing the relevant incident data (based on the desired outcome) using the process described in Chapter 12. A report should be produced describing the analyses, including an Accimap of the contributory factors involved in the incidents.

13.2.3 Step 3: Engage Stakeholders

The third step is engaging relevant stakeholders to participate in design workshops. The workshops work best with 1 facilitator to every 5 participants, they become unwieldy with more than 20 participants in a single workshop. If more stakeholders wish to be involved, then it is usually better to run multiple workshops.

A good starting point for identifying relevant stakeholders is using Rasmussen's (1997) Actormap to map out the various actors and organizations at each level of the system (see Figure 13.2 for an example). Although it may not be possible to engage every actor or organization, the design workshops should include representatives from at least three or four levels of the system. This helps to increase stakeholders' understanding of how the whole system functions, and identify potential unintended consequences of making changes to the system (Rollenhagen, 2011).

The desired outcome (defined in Step 1) is also useful for identifying relevant stakeholders. For example, if the desired outcome is to reduce the injuries associated with a specific work activity, then the design workshops should include the stakeholders who perform the activity, supervise and manage the activity, and impact upon that activity through work scheduling, and designing policy, procedures, and guidelines, and setting budgets.

A 'snowball' strategy is often useful for engaging appropriate stakeholders at the higher levels of the system. This involves asking stakeholders, who have already agreed to participate, to invite their contacts at the higher levels.

13.2.4 Step 4: Workshop Process

The next step is conducting the design workshops. The report on the incident data should be provided to participants well in advance of the workshops to give participants time to digest the information.

The workshops begin by presenting an overview of the design process and the desired outcome, followed by an overview of the systems thinking approach. Participants then form small groups, representing different levels of the system, and follow the design cycle shown in Figure 13.1. A facilitator for each group takes notes throughout the process; it is also useful to record the discussions in each group to provide a way of clarifying strategies at a later date.

The **purge** at the beginning of the design cycle in Figure 13.1 provides an opportunity for participants to express ideas that they may already have about how to "fix" the problem you are going to work on. These may be based on knowledge of the incident data, or just prior assumptions and biases.

To promote creativity and open group discussion, the workshop groups then do a lateral thinking exercise (as shown in Figure 13.1). These are hypothetical situations where you are given little information and must generate possible solutions. For example: solve global warming by 9am tomorrow; play in the World Cup Final this year; end all poverty today; or populate Mars next week. Introducing such exercises forces the brain to think laterally, and promotes similar thinking during the remainder of the session (Read et al., 2017).

Participants then **talk about the analysis** (Figure 13.1), and identify key findings from the report in relation to the desired outcome of the process. The facilitator for each group should be familiar with the data and report, so they can answer any questions. The workshop participants will likely identify additional information about incident causation during this stage, which can be recorded onto the Accimap analyses.

The facilitator then asks the group to **identify goals for incident prevention**, based on the analyses. The goals should be based on an overall problem that is evident from the analysis, rather than focusing on a specific contributory factor in isolation. For example, if many incidents occur during night shifts, and involve issues related to poor decision making and situation awareness, then the overall problem may be fatigue management.

The group then selects a goal to focus on, and develops a **Preventimap**. This involves using the Accimap framework to map out:

1. Incident prevention strategies at each level of the system to address the goal;
2. Supporting strategies required to ensure that the primary strategies are implemented; and
3. The relationships between them.

Each strategy should be described in terms of the actors primarily responsible for implementation, and the specific actions required (e.g. 'Senior executives: Set key performance indicators for middle managers on staff utilizing recreation leave').

During the development of the Preventimap, the facilitator uses the reflection prompts shown in Table 13.1 to help participants generate new strategies, and refocus discussions if participants begin blaming individuals, or focusing on isolated components. The prompts are based on Rasmussen's (1997) risk management framework. Using these prompts will help ensure that the resulting incident prevention strategies are consistent with the systems thinking approach.

A typical workshop involves two to three design cycles. Each design cycle should be followed by a break, and the formation of new groups.

13.2.5 Limitations of the Design Process

This process does not involve any evaluation, or prioritization, of the goals or incident prevention strategies. It is recommended that this takes place after the initial design process, so that the creativity of participants is maximized during the workshops. After the design workshop, the goals for incident prevention should be prioritized based on the number contributory factors and relationships they address from the analysis. Preventimaps can

TABLE 13.1

Reflection Prompts for the Incident Strategy Design Process Based on Rasmussen's (1997) Tenets of Incident Causation

Tenet	Reflection Points
Safety is impacted by the decisions of all actors within the system.	How does the strategy support interaction/coordination across actors at different levels?
Incidents are caused by multiple contributing factors, not a single catastrophic decision or action.	What is needed at the level above to make this strategy work? What is needed at the level below?
Incidents can result from a lack of vertical integration across levels, not just deficiencies at any one level alone.	Can we improve communication and coordination across the levels? Could information flow within actors at the same level be improved?
Actors cannot see how their decisions interact with those made by actors at other levels, so threats to safety are not obvious before an incident.	To make this strategy work what information needs to be communicated up to the higher levels? What would need to be communicated down to the lower levels? To make this strategy work, how does information need to flow between actors – upwards, downwards and within levels of the system? Can we improve feedback across levels of the system so that an actor knows the outcomes of their decisions and actions?
Work practices are not static, they migrate over time due to financial and psychological pressures.	How might financial pressures impact on this strategy, especially over time? Is it financially sustainable? Can we improve this? How might psychological pressures impact on this strategy, especially over time? Will people see its ongoing relevance? Can we improve this?
Migration of work practices can occur at multiple levels, not just in one level alone.	How might financial pressures at a higher/lower level of the system impact on this solution? How might psychological pressures at a higher/lower level of the system impact on this solution?
Migration of work practices causes the system's defences to degrade and erode gradually over time.	How could we monitor the implementation of this strategy over time?

Source: Adapted from Goode, N., Read, G. J. M., van Mulken, M. R. H., Clacy, A., & Salmon, P. M. 2016. *Frontiers in Psychology*, 7.

also be evaluated holistically based on the extent to which they address Rasmussen's tenets (shown in Table 13.1). See Goode et al. (2016) for further details.

13.3 UPLOADS Incident Prevention Strategy Design Process

This section describes the outputs from the incident strategy prevention design process, as applied during the UPLOADS Project.

13.3.1 Desired Outcome of the Design Process

The desired outcome of the process was a reduction of injuries within the LOA sector.

13.3.2 Report on the Incident Data

A report was produced on the injury data collected via UPLOADS from June 2014 to 31st May 2015. The analysis is described in Chapter 12.

13.3.3 Participants

Participants in the workshops included representatives from across the Australian LOA system, as shown in Figure 13.2. In total, 30 people attended two workshops (Workshop 1 = 20, Workshop 2 = 10). On average, participants had 21 years of experience in the LOA sector (SD = 9.52).

13.3.4 Workshop Materials

The following materials were produced for the workshops:

- A report summarizing the analysis of the injury data presented in Chapter 12;
- A workshop agenda;
- A PowerPoint presentation describing the aims of the workshop, with 10 minutes of training on how the systems thinking approach applies to incident prevention (all participants had attended previous workshops on systems thinking, or been involved in collecting data using UPLOADS, so they were all familiar with the approach);
- Large printouts of the UPLOADS injury data Accimap for workshop participants to make notes on;

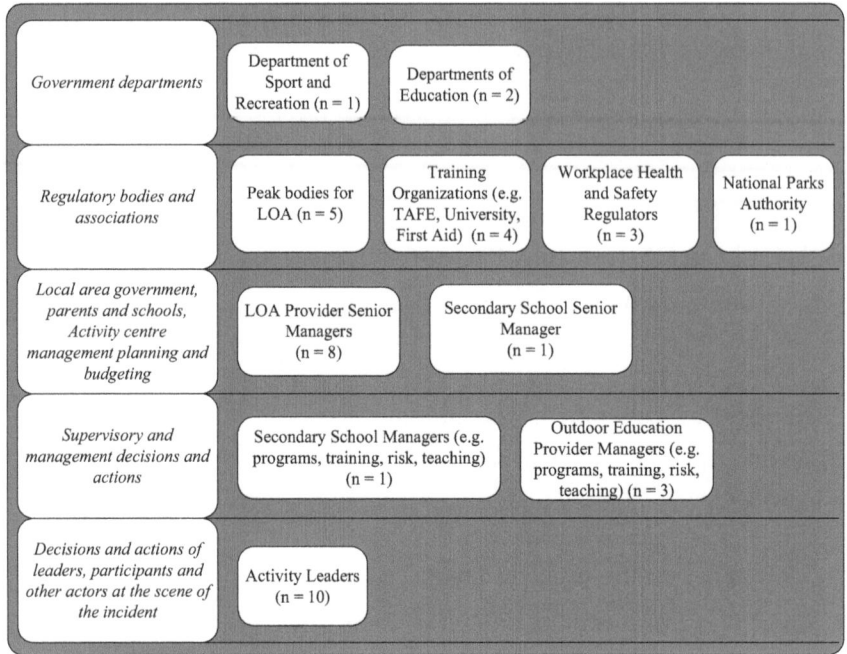

FIGURE 13.2
Number of workshop participants representing actors at each level of the Australian LOA system. (From Goode, N., Read, G. J. M., van Mulken, M. R. H., Clacy, A., & Salmon, P. M. 2016. *Frontiers in Psychology, 7.*)

- Blank Preventimap templates representing the levels of the LOA system; and
- Reflection prompts for facilitators to use to guide discussions (see Table 13.1).

Prior to the workshops, participants were emailed the aims of the workshop, and the report on the analysis. The report was provided to participants to given them ample time to read through the analysis.

13.3.5 The Design Workshops

The design workshops followed the process described in Figure 13.1, and involved three cycles.

The small group discussions were facilitated by five researchers who had been involved in analyzing the data, and writing the report. The workshop discussions were recorded using Dictaphones.

The Preventimap templates were used by the facilitators to record the incident prevention strategies, and the relationships between them, identified during the group discussions. Additional notes were taken by the workshop facilitators to support the later interpretation of the Preventimaps.

13.3.6 Outputs from the Design Process

Based on the report, the workshop groups identified nine specific goals for incident prevention (see Table 13.2). The following sections present some examples of the Preventimaps developed to address these goals, along with a summary of the corresponding workshop discussions.

13.3.6.1 Prevent Burns to Activity Participants during Campcraft

Figure 13.3 shows the Preventimap developed to prevent burns to activity participants during campcraft (i.e. campfire cooking) activities. This goal was identified in response to the finding that campcraft activities had one of the highest rates of injury. Workshop participants considered that many of these injuries occur because: (1) campfire cooking is usually not treated as a separate activity and therefore not risk assessed; (2) cooking is typically perceived as "time off" for the activity leaders and other supervising teachers; (3) cooking and nutrition are not well integrated into the school curriculum; and (4) the LOA regulatory standards do not provide guidance on campcraft activities.

The strategies identified by workshop participants, shown in Figure 13.3, focus on the development and effective implementation of regulatory standards for campcraft activities. The workshop participants proposed that these standards could be built into the existing Adventure Activity Standards (AAS), and the guidelines for the provision of LOA provided by the Department of Education in Victoria.

Several strategies were identified to support the implementation of the new standards, including their inclusion in the curriculum for training activity

TABLE 13.2

Specific Goals for Incident Prevention Identified by Each Workshop Group

Group	Goal
1	Reduce activity leader fatigue during programs
2	Prevent burns to activity participants during campcraft (i.e. campfire cooking)
3	Improve participants skills for outdoor activities
3	Improve the reporting pre-existing injuries
4	Ensure programs are adapted to suit activity participants' competence
5	Improve communication around activity participants' competence levels
5	Improve activity participants' physical literacy
6	Support activity leaders' capacity to undertake dynamic risk assessment
6	Professionalize the career pathway for activity leaders in the LOA sector

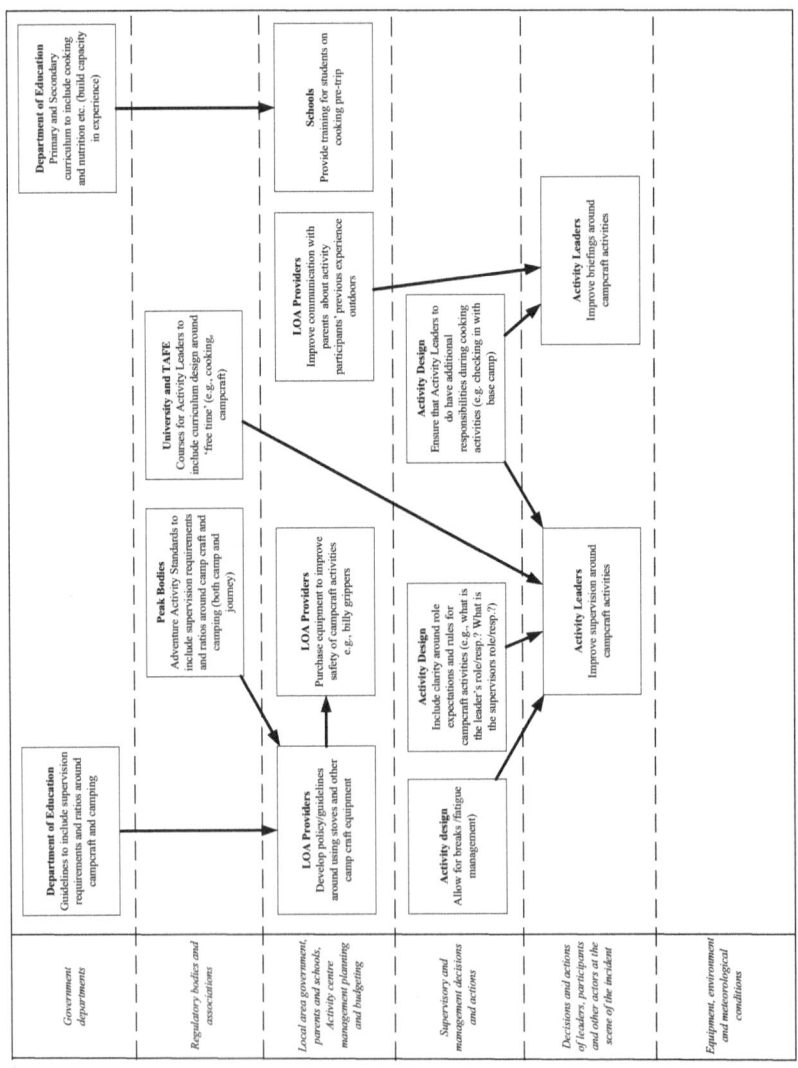

FIGURE 13.3
Preventimap developed to prevent burns to activity participants during campcraft (i.e. campfire cooking) activities.

leaders, the development of appropriate policies and guidelines by LOA providers, and specific changes to program design. Workshop participants also proposed that schools could provide cooking skills training for activity participants prior to LOA programs to prepare them to participate in campcraft activities.

13.3.6.2 Improve Reporting of Activity Participants' Pre-Existing Injuries

Figure 13.4 shows the Preventimap developed to improve the reporting of activity participants' pre-existing injuries prior to LOA programs, so programs and activities can be tailored to prevent further injury.

This goal was identified in response to the finding that activity participants' mental and physical condition played a role 17% of injury incidents;

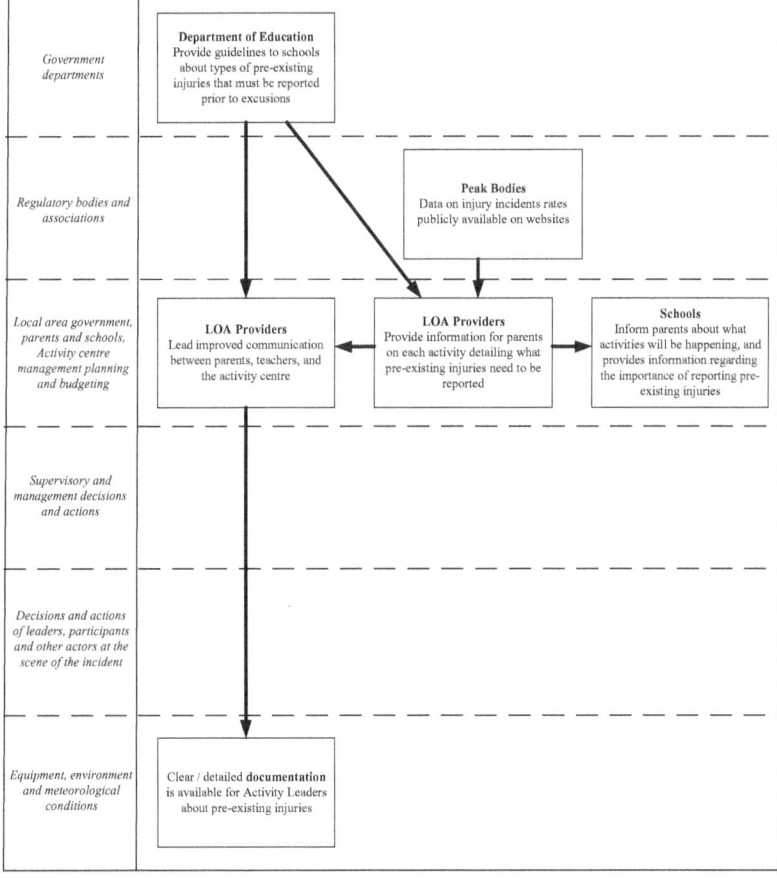

FIGURE 13.4
Preventimap developed to improve the reporting of activity participants' pre-existing injuries prior to LOA programs.

these incidents primarily involved pre-existing injuries. Workshop partici-pants considered that LOA providers were often not provided with accurate information about activity participants' pre-existing injuries.

The strategies identified by workshop participants, shown in Figure 13.4, focus on improving the communication between LOA providers, schools, and parents/carers regarding pre-existing injuries, and ensuring that accu-rate information is available to activity leaders during programs. For exam-ple, workshop participants suggested that the Department of Education could provide guidelines to schools and parents regarding the types of pre-existing injuries that should be reported prior to going on school camps.

13.3.6.3 Ensuring That the Difficulty of LOA Programs Match Participants' Competence Level

Figure 13.5 shows the Preventimap developed by workshop participants to ensure that the difficulty of LOA programs match participants' skill levels. This goal was identified in response to a high frequency of contributory fac-tors relating to activity participants' experience and competence, and com-munication and following instructions. These factors were identified in 24% and 15% of the injury incidents, respectively, and were highly intercon-nected to other factors on the Accimap. Workshop participants considered that many injuries occurred because the design of LOA programs did not adequately consider activity participants' level of experience in the outdoors, and that many activity participants were ill prepared for the programs in terms of skills, fitness, and equipment.

The strategies identified by workshop participants, shown in Figure 13.5, focus on improving communication between different actors regarding partic-ipants' skills, and implementing systems to increase the flexibility of program design. For example, workshop participants suggested that the Department of Education should provide more resources and time to enable schools to pre-pare school students for LOA programs, and gather information about their skills, and abilities, which in turn, would enable schools to collect and pro-vide information to LOA providers on participants' competence. LOA provid-ers could then feed this information down into the development of programs. Workshop participants also suggested that activity leaders needed training on how to identify the skills of participants and adapt programs, as well as specific policies enabling flexibility in program delivery.

13.3.6.4 Supporting Activity Leaders' Capacity to Undertake Dynamic Risk Assessment

Figure 13.6 shows the Preventimap developed by workshop participants to support activity leaders' capacity to dynamically assess changes in environ-mental conditions, and adapt programs accordingly. This goal was identified in response to contributory factors relating to activity leader judgement and

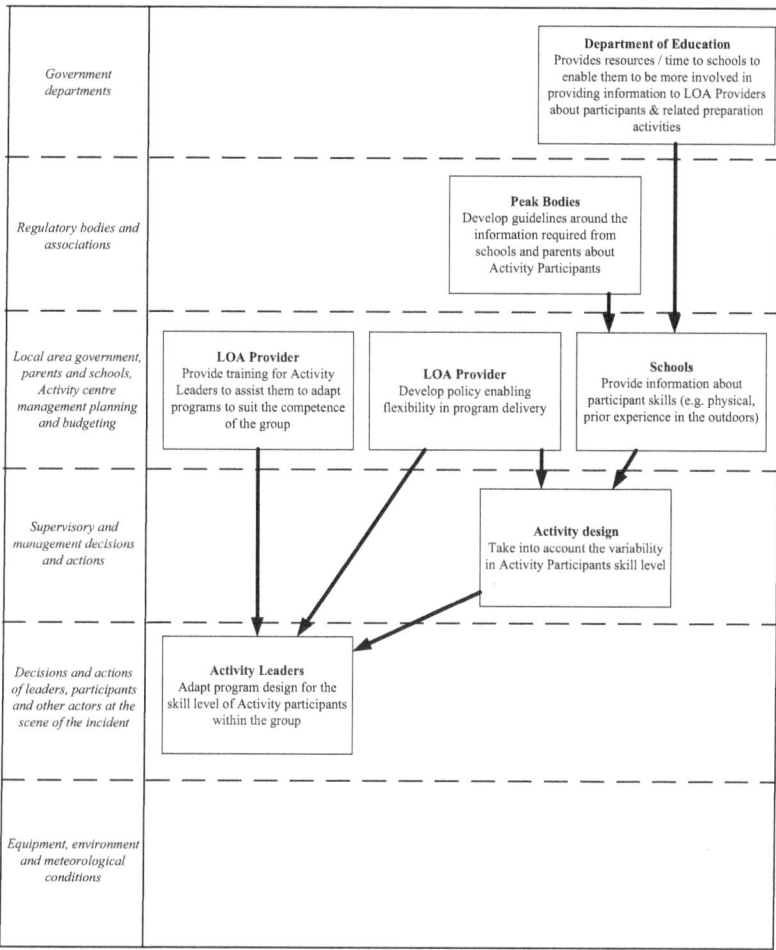

FIGURE 13.5
Preventimap developed to ensure that the difficulty of program matches participants' competence level.

decision–making (2% injury incidents) and situation awareness (1% of injury incidents), which primarily related to failures in identifying changing conditions. Workshop participants discussed the commercial reality of relying on activity leaders with minimal training, who were causal staff. In addition, workshops participants said that LOA providers' policies around dynamic risk assessment, and adapting programs, were often unclear.

The strategies, shown in Figure 13.6, focus on providing training for activity leaders on dynamic risk assessment. In addition to providing training, workshop participants identified several strategies to support its consistent implementation across all LOA providers, including bench marks set by peak bodies, and reducing the reliance on causal staff.

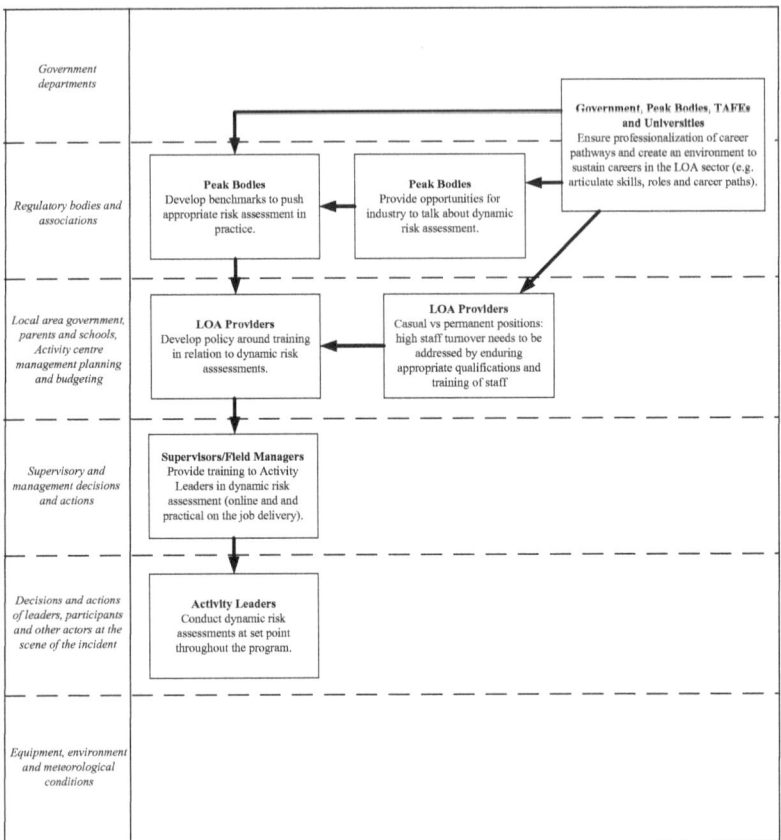

FIGURE 13.6
Preventimap developed to improve activity leaders' capacity to undertake dynamic risk
assessment.

13.3.6.5 *Professionalization of the Career Pathway for Management of LOA Providers*

Figure 13.7 shows the Preventimap developed to improve the career pathway
for managers within LOA providers. This goal was identified in response to a
cluster of contributory factors relating to the management of LOA providers,
including training and evaluation of staff (3% of injury incidents); risk assessment and management (1% of injury incidents); policies and procedures for
activities and emergencies (2% of injury incidents); financial constraints (>1% of
injury incidents); and the supervision and oversight of activities and programs
(>1% of injury incidents). Although these factors were identified in few incidents,
workshop participants considered that they were important, and reflected LOA
providers capacity to train, and retain, high-quality managerial staff.

The strategies shown in Figure 13.7, focus on establishing the LOA sector
as a more professional and prestigious career path by ensuring high-quality

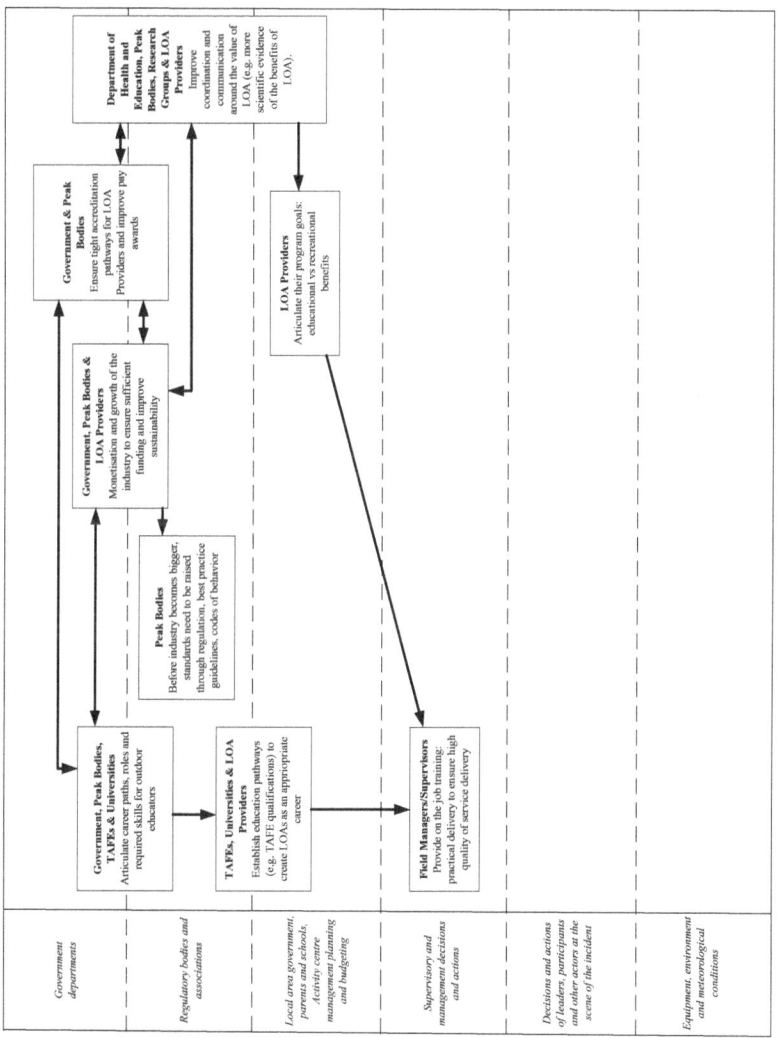

FIGURE 13.7
Preventimap developed to stimulate professionalization of the LOA sector.

qualifications, ongoing professional development and appropriate pay rates. Workshop participants proposed that peak bodies, and government, should tighten accreditation standards, and improve pay awards. The also proposed that government, peak bodies, and LOA providers need to collaborate to ensure monetization and growth of the industry, to ensure sufficient funding and improve sustainability.

13.4 Summary and Conclusions

This chapter has described a process for translating incident analyses into prevention strategies aligned with systems thinking. The process involves stakeholders from across the levels of the system working together to identify goals for incident prevention, and then developing Preventimaps to address these goals. Step-by-step guidance has been provided on how to implement this process in practice. The outputs from the UPLOADS design workshops illustrate that the process is successful in shifting the focus from individual contributory factors at the lower levels of the system, and onto the interactions across the system. We recommend that incident prevention strategy design processes in other safety-critical domains take a similar approach.

References

Dekker, S. (2011). *Drift into failure: From hunting broken components to understanding complex systems.* UK: Ashgate Publishing.

Goode, N., Read, G. J. M., van Mulken, M. R. H., Clacy, A., & Salmon, P. M. (2016). Designing system reforms: Using a systems approach to translate incident analyses into prevention strategies. *Frontiers in Psychology, 7.*

Hollnagel, E. (2012). *FRAM: The functional resonance analysis method: modelling complex socio-technical systems.* UK: Ashgate Publishing.

Leveson, N. (2011). Applying systems thinking to analyze and learn from events. *Safety Science, 49*(1), 55–64.

Rasmussen, J. (1997). Risk management in a dynamic society: A modelling problem. *Safety Science, 27*(2/3), 183–213.

Read, G. J. M., Beanland, V., Lenné, M. G., Stanton, N. A., & Salmon, P. M. (2017). *Integrating human factors methods and systems thinking for transport analysis and design.* Boca Raton, FL: CRC Press.

Rollenhagen, C. (2011). Event investigations at nuclear power plants in Sweden: Reflections about a method and some associated practices. *Safety Science, 49*(1), 21–26. doi:http://dx.doi.org/10.1016/j.ssci.2009.12.012

Vicente, K. J. (1999). *Cognitive work analysis: toward safe, productive, and healthy computer-based work.* Boca Raton, FL: CRC Press.

14

Lessons Learned, Future Research Directions, and the Incident Reporting Systems of Tomorrow

---

Practitioner Summary

The aim of this book was to provide guidance to support the development, testing, and implementation of new incident reporting systems. In this final chapter, we discuss some of the key lessons that we learned through the design and implementation of UPLOADS, along with some critical future research directions for incident reporting systems more generally. We close by discussing some of the key implications of the changing nature of work for the incident reporting systems of the future.

14.1 Introduction

The need for incident reporting systems is clear. Whilst proactive safety management activities are important, it is also critical to learn from past near misses and adverse events (Jacobsson, Ek, & Akselsson, 2011, 2012; Lindberg, Hansson, & Rollenhagen, 2010; Reason, 1997). Although incident reporting systems are now widespread across the safety critical domains, and indeed required by most national and international safety management system standards, many of them are not consistent with contemporary accident causation models, and do not use state-of-the-art analysis methods.

This book has described a process for designing, implementing, and testing new incident reporting systems that is consistent with the principles of Rasmussen's (1997) risk management framework, arguably the most popular and contemporary model of accident causation (Salmon, Goode et al., 2017; Waterson, Jenkins, Salmon, & Underwood, 2017). The intention was to provide guidance to support the development, testing, and implementation of new incident reporting systems, both in domains where incident reporting is already a mainstay (e.g. aviation, healthcare, process control, mining), and in those domains where it is yet to catch on (e.g. sport, peer-to-peer transportation).

We now turn our attention to some of the key lessons that we learned through the design and implementation of UPLOADS, along with some critical future research directions for incident reporting systems more generally. The lessons learned represent key considerations for reader developing and implementing their own systems, whilst the future research directions represent work that needs to be done to ensure that incident reporting systems realize their full potential. Finally, given the fast pace of technological change, it would be foolish to think that incident reporting system requirements and capabilities will remain the same. Accordingly, we close by discussing some of the implications of the changing nature of work for the incident reporting systems of the future.

14.2 Key Findings and Lessons Learned

Through the process of designing and implementing UPLOADS, there were many insights, and lessons learned.

14.2.1 All Systems Go!

Perhaps the first, and most important finding, is that it is possible to develop an incident reporting system that is consistent with the key principles of systems thinking. At the outset of this research program, to our knowledge there were no such systems used in practice. There were doubts as to whether an incident reporting system could be developed, and whether incident reporters would (a) be capable of identifying networks of contributory factors; and (b) feel comfortable enough to report systemic issues relating to management, regulation, and government.

As shown in Chapter 12 and publications (Salmon, Goode et al., 2017), UPLOADS achieves this, and provides the LOA sector with the capacity to report data that sheds light on the network of contributing factors underpinning LOA incidents. Most importantly, the data illustrates that incident reporters are able to identify contributory factors across most levels of the LOA system, and to specify relationships between them. This provides rich data on incidents that can be analyzed in a manner that is consistent with the systems thinking perspective on incident causation. We strongly recommend that incident reporting systems in other safety critical systems are designed, or modified to follow suit. It is our view that the potential of incident reporting systems will not be fulfilled until they are fully aligned with the systems thinking perspective.

14.2.2 A Participatory Design Approach

The importance of involving end users in the design, testing, and implementation of incident reporting systems is demonstrated throughout this book. End users were involved in developing the key components of UPLOADS, including the incident reporting form, software tool, the accident analysis method, and the classification scheme. Likewise, end users formally tested each of these components. Finally, following the analysis of the data reported via UPLOADS, end users designed incident prevention strategies (Chapter 13).

The importance of end-user involvement throughout the entire process cannot be understated, and readers are urged to involve end users in all stages of their own work. We recommend that end users from across the overall system of work are included – this includes the reporters and safety managers who will use the software tool itself, and also those who will access and use incident reporting system outputs (e.g. managers and regulators).

We also recommend working out the best methods for engaging with end users early in the design process. In the LOA sector, phone calls, face-to-face meetings, and the development of personal relationships were highly valued. Many LOA practitioners do not check their emails on a regular basis (if at all). While other sectors may not have the same values, they will have cultural norms around communication. Formally adopting the same practices will help you engage end users in the design process.

14.2.3 Test and Refine, Test and Refine, Test and Refine . . . .

The importance of adopting an iterative design approach, incorporating multiple opportunities to test, and refine, incident reporting system components should also be emphasized. Testing and refinement formed a critical component of the UPLOADS development, and steps were taken to test and refine wherever possible. This involved formal reliability and validity testing, usability testing, a 6-month implementation trial, and finally a national 12-month implementation trial. This testing played a critical role in identifying issues that, if left unresolved, would have adversely impacted the quality of the data and analyses produced, or the uptake of UPLOADS by the sector.

Whilst iterative design reflects good practice, resource constraints often make it difficult to incorporate testing and refinement throughout system design life cycles. We recommend that testing and refinement is undertaken as early as possible, when design concepts are in prototype form, and testing and refinement requires less resources and design changes incur less cost. For example, testing contributory factor classification schemes should occur before the scheme is embedded within any software tool. Likewise, testing of the reporting process should occur when the reporting form is at a paper-based prototype.

14.2.4 Design for Practice, Not for Research

The so-called research-practice gap in human factors and safety science was on our minds throughout the design, testing, and implementation of UPLOADS. This gap is characterized by discrepancies in the methods that safety science researchers and practitioners are applying in response to the same issues. The various constraints that limit the ability of practitioners to use relevant theories and methods are well known. The methods are often overly complex and hard to learn to apply, and they require a considerable time investment to produce a single analysis (Shorrock & Williams, 2016). There is also often a belief that new methods are not required if current methods 'get the job done'.

These constraints were considered during the design and testing of UPLOADS, and a number of the strategies recommended by Shorrock and Williams (2016) were adopted. These included involving practitioners in the methodological development and testing processes, identifying end-user requirements, and developing a 'lite' version of Accimap to incur less resource usage. We recommend that incident reporting system designers adopt similar practices to ensure that they develop systems that are able to be used in practice.

14.2.5 Educating End Users

A final important lesson learned is the requirement to **continually** educate end users on the theoretical underpinnings of the incident reporting system (in this case, systems thinking), and on the art of reporting well. While a proactive program of end-user education was undertaken (i.e. workshops, face-to-face training, presentations, blog pieces, websites, reports), the analyses presented in Chapter 12 showed signs that not all end users understood the systems thinking approach, or how to report incidents in line with it. Just over half of the injury reports contained sufficient information to enable further analysis, and subsequent identification of contributory factors. Further, only 55 reports contained details regarding relationships between factors. This demonstrates the challenges of collecting sufficiently detailed incident reports to support systems analyses. There was also a tendency for reporters to focus on factors related to activity participants, leaders, equipment, and the environment, rather than factors at the higher levels of the framework. One of the reasons for this may be the limited exposure that the sector has previously had to systems thinking. It may also be related to a lack of awareness of how higher-level factors, such as actions and decisions by the government or regulatory bodies, impact the delivery of LOAs. Further education across the LOA sector is now required on: (a) the need for managers to review reports; (b) the systems thinking approach; and (c) how factors across the LOA system might influence participant and leader decision making and behavior.

14.3 Further Research on Incident Reporting Systems

We hope that this book narrows the gap between research and practice in the area of incident reporting systems. It is important to note, however, that improving the design of incident reporting systems is merely the first step in enhancing their utility for safety management, and there are some critical issues that require more research in this area.

14.3.1 Using Incident Reporting System Outputs Effectively

The point of reporting is to help prevent recurrence of similar incidents. Reporting is only the first step in a sequence known as the learning cycle that also includes analysis, decision making, implementation, and follow-up (Jacobsson et al., 2012). Unfortunately, many incident reporting systems are not embedded in a formal learning cycle, as we propose in Chapter 3. In particular, there is little evidence that the data collected by incident reporting systems is actually being used to inform the development of incident prevention strategies (Benn et al., 2009; Drupsteen & Hasle, 2014; Pham, Girard, & Pronovost, 2013; Tighe, Woloshynowych, Brown, Wears, & Vincent, 2006).

Whilst the process described in Chapter 13 provides an approach for designing strategies that are consistent with systems thinking, it is clear that further work is required to better integrate internal incident reporting systems within formal learning cycles. While UPLOADS has enhanced the quality of incident reporting data and analysis, it has also raised questions about how organizations should respond to the analyses produced.

In particular, it is difficult to prevent organizations from focusing on 'broken components' such as the worker (e.g. retraining) or equipment (e.g. purchase new equipment), despite the fact that other systemic causes are reported. The flaws associated with this approach are well known (Dekker, 2011). Instead, treatment of wider systems factors is more effective than the treatment of components at the 'sharp end' of system operation, since the factors shaping behavior are removed (e.g. Dekker, 2011; Rasmussen, 1997).

Without guidance, however, it is difficult for organizations to concep-
tualize appropriate system reforms based on incident reporting data
(Drupsteen, Groeneweg, & Zwetsloot, 2013; Lundberg, Rollenhagen, &
Hollnagel, 2009; Rollenhagen, Westerlund, Lundberg, & Hollnagel, 2010).
This lack of guidance for translating incident reporting systems outputs
into prevention strategies, coupled with the inherent difficulty in developing strategies in line with the systems thinking philosophy, is limiting
the impact of incident reporting systems. This is not only a problem for
LOA providers; it has been recognized more widely as an international
knowledge gap for all injury problems (Finch, 2011, 2012; Hanson, Finch,
Allegrante, & Sleet, 2012).

In response to this, the authors are currently engaged in a program of
research involving the development of a structured approach for developing
incident prevention strategies, based on the analyses provided by UPLOADS.
Known as UPLOADS PrIMe (Preventing Incidents Method), the process is
designed to ensure that the strategies that organizations produce are aligned
with systems thinking, respond to the network of contributory factors underpinning incidents, and that an appropriate implementation, evaluation, and
feedback process is included. This will ensure that UPLOADS is appropriately embedded in a formal learning cycle. It is recommended that similar
research is undertaken in other areas to enable incident reporting systems to
be embedded in a similar learning cycle.

14.3.2 Testing the Impact of Incident Reporting Systems

Although the use of incident reporting systems is widespread, and many
have discussed their important role within safety management (Jacobsson
et al., 2011, 2012; Lindberg et al., 2010; Reason, 1997), there is in fact little
evidence to demonstrate that they provide a safety benefit in terms of reductions in incidents. Few formal studies have been conducted on the effects of
incident reporting systems on actual safety outcomes (Finch, 2006; Johnston,
2009; Lindberg et al., 2010; Pless, 2008). This is a key need for future research.
As part of the UPLOADS PrIMe research program described above we will
formally test the safety benefits of using UPLOADS. This involves a two year
study examining the impact of UPLOADS, and UPLOADS PrIMe, on safety
and incident prevention strategies. This is by no means a simple undertaking, it will involve collection of data from LOA providers assigned to the
following three groups:

1. LOA providers using UPLOADS alone;
2. LOA providers who have adopted the new incident reporting and
   learning cycle (UPLOADS + UPLOADS PrIMe); and
3. LOA providers their own internal incident reporting system.

A range of data will be collected over the two year study period to support both qualitative and quantitative analyses. This will include incident and participation data, as well as information regarding the incident prevention strategies developed in response to incidents, and the success of their implementation. Analysis of the data will provide evidence on the safety benefits of adopting a systems thinking–based incident reporting and learning cycle. It is recommended that similar studies are undertaken in other domains to assess the impact of incident reporting systems.

14.4 Incident Reporting in 2050

If we believe the hype, the work systems of 2050 will be almost unrecognizable. Robots will replace many human workers, and Artificial General Intelligence (AGI)-based systems will perform most of the work. The role of human workers will become far more supervisory in nature; however, we will need to have the capacity to intervene when our robot and AI counterparts struggle to cope with complex and unexpected scenarios. On top of this, advances in big data-based systems will mean that system performance will be closely scrutinized via the continuous recording, and interrogation, of multiple data sources. Dynamic performance monitoring will allow automation to intervene when things appear to be going wrong. Big Brother will be watching.

It is hard to imagine how this vision would apply to LOAs and the great outdoors; however, it is a likely reality for most of the safety critical systems in which incident reporting systems are currently used. In aviation, for example, automation already does most of the work once aeroplanes are in the sky. By 2050, aircraft will be flown by advanced automation that is capable of making the decisions that pilots currently make. Human supervisors (formerly known as pilots) will be located either in the cockpit, or on the ground, with the ability to take control, but only when necessary.

The implications for incident reporting systems, and indeed incident causation, are profound. Of course, incidents will still occur. Automation and AI will not be the silver bullets that they are portrayed to be, at least not in the short term. When incidents occur, the onus might initially be on non-human agents to identify and report them. Advanced automated systems will be able to provide detailed data on performance parameters as well as other factors that might have played a role in an incident (e.g. weather conditions or maintenance activities). Human supervisors will also contribute to incident reports; however, often they will be detached from the sharp end. Incident reports will comprise a combination of objective and subjective data, but the reports will be heavily based on the former rather than the latter. This is in contrast with most current incident reporting systems which operate primarily on subjective data, with any objective data viewed as a bonus.

The nature of the incidents themselves will necessarily change. Increasingly technology-centric and complex systems will mean that a wider set of contributory factors will play a role in adverse events. For example, there will be issues associated with automation and AI (see the recent Tesla and Uber fatal road crashes), as well as related issues with regulation, design guidelines, standards, and certification processes. As the bureaucracy required to control work systems expands, the network of contributory factors at the higher levels of work systems is likely to become more dense and complex, whilst the contributory factors at the lower levels of work systems will likely shift in nature. The Accimaps of tomorrow will look very different to the Accimaps of today.

To remain useful, incident reporting systems will have to adapt. While the components of incident reporting systems will likely be the same (e.g. incident data collection fields, accident analysis method), the components themselves will need to be different. It is not beyond the realms of possibility to suggest that this book will no longer be at all relevant in 2040. Indeed, we would go as far to say that parts of it will no longer be relevant by 2030. As a result, work is required to ensure that incident reporting systems are able to adapt, and stay relevant.

14.4.1 Systems Thinking and Beyond?

In Chapters 1 and 2 we argued that incident reporting systems should be better aligned with the systems thinking approach. This argument was built on the notion that significant progress has been made by applying systems thinking in incident analysis and prevention activities. The changing nature of work systems, however, means that further work is required. Indeed, although the systems thinking approach has supported great progress in safety performance in many sectors of industry, this progress is now slowing (Salmon, Walker, Read, Goode, & Stanton, 2017; Walker, Salmon, Bedinger, & Stanton, 2017). In many sectors safety improvements are reaching a plateau, or incident rates are increasing (Dekker & Pitzer, 2016; Vincent & Amalberti, 2016; Walker et al., 2017).

Accident theory and analysis methods have been implicated in this slowing of progress (Dekker & Pitzer, 2016). Leveson (2011), for example, argues that our accident analysis methods do not fully uncover the underlying causes of incidents. In Salmon and Walker et al. (2017) we argued that even our state-of-the-art methods do not enable us to fully understand incidents, or how to prevent them. Researchers are also pointing to gaps in contemporary accident causation theory, and a pressing need to update it. Hollnagel (cited in Waterson et al., 2015), for example, argues that most accident analysis methods are dated, and are now unfit for purpose.

Rasmussen's risk management framework and associated Accimap methodology provide a good example. Both were developed well over 20 years ago when safety critical systems, and indeed the kinds of contributory factors involved in incidents, were different. Although Rasmussen's risk management framework pays lip service to various elements of complexity, the Accimap method cannot cope analytically with many of them. For example, Accimap is challenged by features such as emergence, coupling, performance variability, normal performance, and complexity itself (Salmon, Walker et al., 2017; Stanton & Harvey, 2017; Walker et al., 2017). These issues are to be found, increasingly prominently, in incidents, as are new, and emergent, contributory factors, which confound methods underpinned by contributory factor classification schemes.

Theoretical and methodological development is therefore required for accident models and methods to remain fit for purpose (Salmon, Walker et al., 2017). By association, the incident reporting systems of the future should be developed based on advances in accident causation models and methods. Our message to the reader is that, although systems thinking currently provides suitable models and methods to support the development of incident reporting systems, further work is required to ensure that they continue to evolve and remain appropriate. The authors are currently embarking on a major program of work that involves updating models and methods, such as Rasmussen's, to ensure they remain appropriate for the safety critical systems of tomorrow.

14.5 Summary

The aim of this book was to provide a process for designing, testing, and implementing an incident reporting system that is consistent with systems thinking. While the examples used relate to an incident reporting system that was developed for the LOA sector in Australia, the process itself is generic, and can be applied in any safety-critical domain. It is hoped that researchers and practitioners working in the area of incident reporting use the process described (or parts of it). To support this, this chapter outlined some of the key lessons that we learned when going through the process of developing UPLOADS. It is our view that these represent important considerations for the reader when developing and implementing their own incident reporting systems.

We also identified a series of areas for further research that are required for incident reporting systems to realize their full potential. No doubt there are more than those discussed, and we encourage other researchers and practitioners to pursue these ideas, and push the boundaries of incident reporting systems.

14.6 Conclusions

To close the book, it is worth reiterating the important role that incident reporting systems play in safety management. While proactive safety management is important, it is also critical to learn from the adverse events and near misses. In his seminal book *The Human Contribution*, James Reason describes incident reporting as an essential prerequisite to managing safety. He describes the idea of collective mindfulness as a critical component of organizational safety and resilience. According to Reason (2008), collectively mindful organizations:

- Work hard to extract high value from incident data;
- Actively create a reporting culture;
- View failures or errors as emerging from the interactions between many contributory factors;
- Strive for system reforms rather than local repairs;
- Are aware that failure can take a variety of yet-to-be-encountered forms;
- Continually look for new ways in which contributory factors can interact; and
- Are pre-occupied with failure.

By designing and implementing incident reporting systems based on the guidance in our book, the reader will be creating the conditions for collective mindfulness in their own organizations. We look forward to seeing the results!

References

Benn, J., Koutantji, M., Wallace, L., Spurgeon, P., Rejman, M., Healey, A., & Vincent, C. (2009). Feedback from incident reporting: Information and action to improve patient safety. *BMJ Quality & Safety, 18*(1), 11–21. doi:10.1136/qshc.2008.031997

Bureau of Infrastructure Transport and Regional Economics. (2017). Road Trauma Australia—Annual Summaries. Retrieved from https://bitre.gov.au/publications /ongoing/road_deaths_australia_annual_summaries.aspx

Dekker, S. (2011). *Drift into failure: From hunting broken components to understanding complex systems*. Aldershot, UK: Ashgate.

Dekker, S., & Pitzer, C. (2016). Examining the asymptote in safety progress: A literature review. *International Journal of Occupational Safety and Ergonomics, 22*(1), 57–65.

Drupsteen, L., Groeneweg, J., & Zwetsloot, G. (2013). Critical steps in learning from incidents: Using learning potential in the process from reporting an incident to accident prevention. *International Journal of Occupational Safety and Ergonomics, 19*(1), 63–77. doi:10.1080/10803548.2013.11076966

Drupsteen, L., & Hasle, P. (2014). Why do organizations not learn from incidents? Bottlenecks, causes and conditions for a failure to effectively learn. *Accident Analysis & Prevention, 72*, 351–358. doi:10.1016/j.aap.2014.07.027

Finch, C. F. (2006). A new framework for research leading to sports injury prevention. *Journal of Science and Medicine in Sport, 9*(1), 3–9.

Finch, C. F. (2011). No longer lost in translation: The art and science of sports injury prevention implementation research. *British Journal of Sports Medicine, 45*, 1253–1257. doi:10.1136/bjsports-2011-090230

Finch, C. F. (2012). Implementing and evaluating interventions. *Injury Research* (pp. 619–639). Springer.

Hanson, D. W., Finch, C. F., Allegrante, J. P., & Sleet, D. (2012). Closing the gap between injury prevention research and community safety promotion practice: Revisiting the public health model. *Public Health Reports, 127*(2), 147–155. doi:10.1177/003335491212700203

Jacobsson, A., Ek, A., & Akselsson, R. (2011). Method for evaluating learning from incidents using the idea of 'level of learning'. *Journal of Loss Prevention in the Process Industries, 24*(4), 333–343. doi:http://dx.doi.org/10.1016/j.jlp.2011.01.011

Jacobsson, A., Ek, A., & Akselsson, R. (2012). Learning from incidents – a method for assessing the effectiveness of the learning cycle. *Journal of Loss Prevention in the Process Industries, 25*(3), 561–570. doi:10.1016/j.jlp.2011.12.013

Johnston, B. D. (2009). Surveillance: To what end? *Injury Prevention, 15*(2), 73–74. doi:10.1136/ip.2009.021790

Leveson, N. G. (2011). Applying systems thinking to analyze and learn from events. *Safety Science, 49*(1), 55–64. doi:10.1016/j.ssci.2009.12.021

Lindberg, A.-K., Hansson, S. O., & Rollenhagen, C. (2010). Learning from accidents – what more do we need to know? *Safety Science, 48*(6), 714–721. doi:10.1016/j.ssci.2010.02.004

Lundberg, J., Rollenhagen, C., & Hollnagel, E. (2009). What-you-look-for-is-what-you-find – the consequences of underlying accident models in eight accident investigation manuals. *Safety science, 47*(10), 1297–1311. doi:10.1016/j.ssci.2009.01.004

Pham, J. C., Girard, T., & Pronovost, P. J. (2013). What to do with healthcare incident reporting systems. *Journal of Public Health Research, 2*(3). doi:10.4081/jphr.2013.e27

Pless, B. (2008). Surveillance alone is not the answer. *Injury Prevention, 14*(4), 220–222. doi:10.1136/ip.2008.019273

Rasmussen, J. (1997). Risk management in a dynamic society: A modelling problem. *Safety Science, 27*(2/3), 183–213.

Reason, J. (1997). *Managing the risks of organizational accidents.* Aldershot, UK: Ashgate.

Reason, J. (2008). *The human contribution: Unsafe acts, accidents and heroic recoveries.* Surrey, UK: Ashgate.

Rollenhagen, C., Westerlund, J., Lundberg, J., & Hollnagel, E. (2010). The context and habits of accident investigation practices: A study of 108 Swedish investigators. *Safety Science, 48*(7), 859–867. doi:http://dx.doi.org/10.1016/j.ssci.2010.04.001

Salmon, P. M., Goode, N., Lenné, M. G., Finch, C. F., & Cassell, E. (2014). Injury causation in the great outdoors: A systems analysis of led outdoor activity injury incidents. *Accident Analysis & Prevention, 63,* 111–120. doi:10.1016/j.aap.2013.10.019

Salmon, P. M., Goode, N., Taylor, N., Lenné, M. G., Dallat, C. E., & Finch, C. F. (2017). Rasmussen's legacy in the great outdoors: A new incident reporting and learning system for led outdoor activities. *Applied Ergonomics, 59,* 637–648. doi:10.1016/j.apergo.2015.07.017

Salmon, P. M., Walker, G. H., Read, G., Goode, N., & Stanton, N. A. (2017). Fitting methods to paradigms: Are ergonomics methods fit for systems thinking? *Ergonomics, 60*(2), 194–205. doi:10.1080/00140139.2015.1103385

Shorrock, S. T., & Williams, C. A. (2016). Human factors and ergonomics methods in practice: Three fundamental constraints. *Theoretical Issues in Ergonomics Science, 17*(5–6), 468–482.

Stanton, N. A., & Harvey, C. (2017). Beyond human error taxonomies in assessment of risk in sociotechnical systems: A new paradigm with the EAST 'broken-links' approach. *Ergonomics, 60*(2), 221–233. doi:10.1080/00140139.2016.1232841

Tighe, C. M., Woloshynowych, M., Brown, R., Wears, B., & Vincent, C. (2006). Incident reporting in one UK accident and emergency department. *Accident and emergency nursing, 14*(1), 27–37. doi:10.1016/j.aaen.2005.10.001

Vincent, C., & Amalberti, R. (2016). *Safer healthcare.* Cham: Springer International Publishing.

Walker, G. H., Salmon, P. M., Bedinger, M., & Stanton, N. A. (2017). Quantum ergonomics: Shifting the paradigm of the systems agenda. *Ergonomics, 60*(2), 157–166. doi:10.1080/00140139.2016.1231840

Waterson, P., Jenkins, D. P., Salmon, P. M., & Underwood, P. (2017). 'Remixing Rasmussen': The evolution of Accimaps within systemic accident analysis. *Applied Ergonomics, 59,* 483–503. doi:10.1016/j.apergo.2016.09.004

Waterson, P., Robertson, M. M., Cooke, N. J., Militello, L., Roth, E., & Stanton, N. A. (2015). Defining the methodological challenges and opportunities for an effective science of sociotechnical systems and safety. *Ergonomics, 58*(4), 565–599. doi 10.1080/00140139.2015.1015622

Appendix A: UPLOADS Contributory Factor Classification Scheme

This section provides an overview of the UPLOADS contributory factor classification scheme, along with the examples that were developed to support the consistent selection of categories within the incident reporting system. The examples were developed based on an analysis of over 1,000 incidents reports, and feedback from analysts across multiple cycles of reliability testing.

A.1 Structure of the Classification Scheme

The classification scheme consists of two levels of categories:

- Level 1: The led outdoor activity system
- Level 2: Descriptive categories

Each level is described in detail in the following sections.

A.2 Level 1: The Led Outdoor Activity System

The first level describes the actors, artefacts, and activity context within the led outdoor activity system in terms of 14 broad categories:

1. **Activity Equipment and Resources:** The artefacts required to conduct the activity.
2. **Activity Environment:** The immediate environment in which the activity takes place.
3. **Activity Leader:** The people instructing the activity (e.g. leaders, guides, and instructors).

4. **Activity Participants:** The people actively participating in the activity (e.g. students or clients).

5. **Other People in Activity Group:** The people in the activity group who are not participating in the activity but contribute to the immediate supervision and provision of the activity (e.g. parents, teachers, drivers, cooks, and cleaners).

6. **Activity Group:** The dynamics within the group and factors impacting on group performance.

7. **Other People in Activity Environment:** The people in the immediate activity environment who are not part of the activity group (e.g. members of the public or emergency services).

8. **Supervisor/Field Managers:** The people who contribute to the planning/supervision of the activity, and supervision of activity leaders. These people are typically external to the context of the activity environment. This category includes field managers, supervisors, and administrative staff.

9. **Higher-Level Management:** The people who contribute to the strategic management of the activity and program (e.g. activity centre managers, senior managers, and CEOs).

10. **Local Area Government:** The local area government where the activity takes place.

11. **Schools:** Schools and school staff with students involved in activities and programs.

12. **Parents/Carers:** Parents and carers who have children involved in activities and programs.

13. **Regulatory Bodies and Professional Associations:** Regulatory bodies and professional associations relevant to led outdoor activity provision, including peak bodies, and education and training providers.

14. **State and Federal Government:** Government departments and bodies relating to the provision of led outdoor activities (e.g. National Parks and Wildlife Service; Sport and Recreation State Government departments; and the Australian Sports Commission).

A.3 Level 2: Descriptive Categories

The second level breaks the first level categories down into more descriptive categories, with extensive examples of the types of contributory factors that should be classified within each category.

A.3.1 Activity Equipment and Resources

Level 2 Categories	Examples
Documentation	Lack of/incorrect maps; participant lists; participant details; consent forms.
Equipment, clothing, and personal protective equipment	Missing/inappropriate/broken paddles; helmets; boots; jackets; incorrect use of equipment; failure to use equipment.
Food and drink	Inadequate water; not eating enough; spoiled or uncooked meat.
Medication (for those involved in the activity)	Missing/wrong asthma medication.
Other	Activity equipment and resources factors not otherwise classified.

A.3.2 Activity Environment

Level 2 Categories	Examples
Animal and insect hazards	Animal attacks; insect bites.
Infrastructure and terrain	Maintenance of activity area; huts; rocky trails; slippery terrain.
Trees and vegetation	Falling branches; thorns; stinging nettles.
Water conditions	Fast flowing water; rapids; swell; current.
Weather conditions	Excessively cold/hot conditions; rain; snow; visibility; storms.
Other	Activity environment factors not otherwise classified.

A.3.3 Activity Leader

Level 2 Categories	Examples
Communication, instruction, and demonstration	Lack of communication with participants; teachers, etc.; communication of wrong information; missing critical elements from instruction; didn't demonstrate safety procedure; didn't provide training in activity.
Compliance with procedures, violations, and unsafe acts	Didn't comply with company policies regarding safety equipment; horsing around in dangerous area.
Experience, qualifications, competence	Unfamiliar with location/activity; lack of qualifications for activity; lack of competence in activity.
Judgement and decision making	Poor judgement regarding participant's ability; failure to make a decision to cancel activity due to conditions.
Mental and physical condition	Fatigue; injury; including pre-existing injuries; panic; intoxicated.
Planning and preparation	Failure to check equipment; did not consider all risks/hazards in activity planning.
Situation awareness	Poor understanding of environment; unaware of important information; inattention; hazard awareness; lack of understanding of 'what is going on'.
Supervision/leadership of activity	Not supervising participants.
Other	Activity leader factors not otherwise classified.

A.3.4 Activity Participant

Level 2 Categories	Examples
Communication and following instructions	Poor communication with activity leader; teachers; ignoring instructions.
Compliance with procedures, violations, and unsafe acts	Didn't comply with safety procedures; walking into fenced off/unauthorized area; horsing around in dangerous area.
Experience and competence	Lack of skills; inexperienced in the activity in question.
Judgement and decision making	Descended too early; decided not to tell the leader they felt sick/injured.
Mental and physical condition	Fatigue; injury; including pre-existing injuries; panic; intoxicated.
Planning and preparation for activity/trip	Did not bring appropriate clothing/equipment; did not complete fitness training; failed to bring medication/water.
Situation awareness	Poor understanding of environment; unaware of important information; inattention; lack of hazard awareness; lack of understanding of 'what is going on'.
Other	Activity participant factors not otherwise classified.

A.3.5 Other People in Activity Group (Not Actively Participating)

Level 2 Categories	Examples
Communication and following instructions	Lack of communication with activity leader, participants, teachers, supervisors; communication of wrong information; ignoring instructions.
Compliance with procedures, violations, and unsafe acts	Didn't comply with safety procedures; walking into fenced off area; horsing around in dangerous area.
Experience, qualifications, competence	Lacked the experience to assist in supervising activity, inexperienced in activity in question.
Judgement and decision making	Did not take participant injury/complaint seriously.
Mental and physical condition	Fatigue; Injury, including pre-existing injury; panic; intoxicated.
Planning and preparation for activity/trip	Did not bring appropriate clothing/equipment/participant documentation.
Situation awareness	Poor understanding of environment, unaware of important information; inattention, hazard awareness, lack of understanding of 'what is going on'.
Supervision of activity	Not supervising participants.
Other	Other people in activity group factors not otherwise classified.

A.3.6 Activity Group

Level 2 Categories	Examples
Communication within group	Loss of contact between group members; failure of participants to communicate with one another.
Group composition	Variable abilities; different ages.
Group dynamics	Bullying; competition within group; peer pressure; morale.
Group size	Group was too large for number of activity leaders/activity.
Late arrival of group	Group arrived 2 hours late to activity centre.
Teamwork	Division in group; not working together.
Time pressure	Not enough time to complete activity safely due to planning.
Other	Activity group factors not otherwise classified.

A.3.7 Other People in Activity Environment (Not in Activity Group)

Level 2 Categories	Examples
Communication	Gave wrong directions; ambulance officer could not find group.
Compliance with procedures, violations, and unsafe acts	Lighting campfire during total fire ban; trespassing on activity centre property.
Experience, qualifications, competence	Member of the public lacked skills to complete climb and activity leader had to help them.
Judgement and decision making	Member of the public gave warning that activity was dangerous.
Mental and physical condition	Member of the public fatigue; injuries; panic; intoxicated.
Planning and preparation	Member of the public not properly prepared for activity and requiring assistance.
Situation awareness	Poor understanding of environment; unaware of important information; inattention; hazard awareness; lack of understanding of 'what is going on'.
Other	Other people in activity environment factors not otherwise classified.

A.3.8 Supervisors/Field Manager

Level 2 Categories	Examples
Activity or program design	Poorly designed activity; too many activities were scheduled during program; activities were inappropriate for participant skill level.
Communication	Lack of communication with activity leader; participants; teachers; supervisors.
Compliance with procedures, violations and unsafe acts	Did not comply with company regulation regarding group size/activity leader qualifications.
Experience, qualifications, competence	Did not have specific experience in the activity/activity location.
Judgement and decision making	Failure to decide to cancel activity due to conditions.
Mental and physical condition	Fatigue; panic; injuries.
Planning and preparation for activity	Didn't maintain activity equipment; did not obtain consent forms
Supervision of activity leaders and other staff	Leaders are pretty much left to fend for themselves.
Supervision/oversight of programs/activities	Field manager was not contactable during the emergency.
Other	Supervisors/Field manager factors not otherwise classified.

A.3.9 Higher-Level Management

Level 2 Categories	Examples
Communication	Lack of communication with field managers/activity leaders/schools.
Financial constraints	There are no resources to increase staff levels; no budget available for new equipment; poor maintenance procedures.
Judgement and decision making	Failure to decide to cancel activity due to conditions.
Organizational culture	The same type of incidents keeping happening; it is normal to violate this procedure; activities go ahead no matter the weather; high risk is favored by the activity leaders.
Policies and procedures for activities and emergencies	Procedures do not clearly describe the preferred abseiling method; there were no emergency procedures available for this activity; consent forms or policy is inadequate.
Risk assessment and management	Risk assessment wasn't conducted/applied during the planning of an activity.
Staffing and recruitment	Not enough staff have been employed to safely supervise all activities; inappropriately qualified staff were recruited.
Supervision of staff (e.g. activity leaders, field managers)	Field manager's decisions were not monitored.
Supervision/oversight of activities and programs	Activity centre manager was not contactable during the emergency.
Training and evaluation of staff (e.g. activity leaders, field managers)	The training program only covers ideal conditions for the activity; training program inadequate; lack of evaluation of activity leader skills.
Other	Higher-level management factors not otherwise classified.

A.3.10 Local Area Government

Level 2 Categories	Examples
Auditing	Food safety inspector had not visited the catering company; an audit failed to identify any issues related to the activity in question.
Communication	Failure to communicate changes in fire evacuation plans to activity centre.
Funding and budgets	Lack of funding to maintain council campgrounds.
Legal responsibility for safety within the council area	Failure to ensure campgrounds are properly maintained; lack of fences around dangerous areas.
Policies and procedures	Fire evacuation plans did not consider campgrounds.
Other	Local area government factors not otherwise classified.

A.3.11 Schools

Level 2 Categories	Examples
Communication	Did not provide information about participants behavioral issues; injuries; medical conditions to activity centre; did not provide appropriate information about the risks involved in activity to parents.
Dropping off/picking up participants	School group arrived late; transport provided was inappropriate for terrain.
Judgement and decision making	Poor decision that participant was well enough for trip or that participant was capable of undertaking activity.
Legal responsibility for safety of staff and students	Did not ensure activity centre was accredited; approved the conduct of potentially dangerous activity.
Planning and preparation for activity/trip	Did not collect information on students/staff required by activity centre.
Policies and procedures	Policies/procedures regarding outdoor activities are unclear/non-existent.
Teacher/student ratio	Only 2 teachers were sent to supervise 50 students.
Other	Schools factors not otherwise classified.

A.3.12 Parents/Carers

Level 2 Categories	Examples
Communication	Did not provide information about participants behavioral issues; injuries; medical conditions; did not understand the risks associated with activity.
Dropping off/picking up participants	Did not pick up participant on time.
Judgement and decision making	Poor decision that participant was well enough for trip or that participant was capable of undertaking activity.
Legal responsibility for safety of child	Did not ensure activity centre was accredited.
Planning and preparation for activity/trip	Did not buy adequate clothing for participant; provided inadequate clothing.
Other	Parents/Carers factors not otherwise classified.

A.3.13 Regulatory Bodies and Professional Associations

Level 2 Categories	Examples
Accreditation/licensing	Lack of/inappropriate accreditation scheme for activity centres.
Auditing	Audit scheduled but was not complete; an audit failed to identify any issues related to the activity in question.
Communication	Did not communicate changes to activity guidelines to activity centres; engagement with practitioners/experts to set standards.
Curriculum of outdoor education/recreation qualifications	TAFE Cert. IV did not adequately prepare activity leaders to supervise young children.
Funding and budgets	Lack of funding to revise activity guidelines.
Interactions with government	Failure to obtain funding for sector.
Standards and code of practice	Activity standard did not consider sea conditions.
Other	Regulatory bodies and professional associations factors not otherwise classified.

A.3.14 State and Federal Government

Level 2 Categories	Examples
Communication	Government policy regarding shooters use of the national park was not communicated well.
Funding and budgets	Lack of funding for the outdoor activity sector.
Infrastructure and land management	Trail has not been maintained by the government department.
Policies and legislation	The activity is not covered under current legislation.
Other	State and federal government factors not otherwise classified.

Appendix B: Examples of Coding Tasks for Reliability and Validity Assessments

As described in Chapter 7, there are three different ways to present coding tasks during reliability and validity assessments. This chapter presents examples of each approach using the UPLOADS contributory factor classification scheme. Regardless of the type of coding task, participants should be first presented with an incident description (see below), and an overview of the classification scheme with examples of the types of contributory factors that should be classified within each category (as presented in Appendix A).

Example Incident Description

It was the last climb of the day, around 3pm. The group (4 adult participants and 1 activity leader) had been climbing since 10am, with only a short break for lunch. The activity leader was demonstrating how to lead climb the 'Falcon's Nest'. The activity leader reported that at the time he was feeling tired after a really long, hot day, and was looking forward to 'getting it over with'. Although he had worn a helmet throughout the day, he forgot to put it on for this climb (company procedures state that a helmet should be worn at all times during an active climb, even if you are only an observer). A participant was acting as the belayer; he failed to warn the activity leader about the forgotten helmet. The activity leader placed a bolt plate onto the first bolt. He then uncharacteristically clipped a quickdraw to his rope first, and then proceeded to clip the quickdraw (with rope attached) to the bolt plate. He had made one move past the protection when his belayer alerted him that the quick draw had become disconnected from the bolt plate. The activity leader is unsure how this happened. He tried to down climb the move to rectify the problem, but he became anxious as he remembered that he had forgotten his helmet, and he slipped. A 7 m ground fall resulted. The belayer (who was closest to the injured activity leader) attended to the activity leader. No first aid was administered, but it was clear that they would be unable to move the activity leader. One of the participants contacted rescue authorities using a mobile phone. Police rescue, ambulance, and SES volunteers attended the casualty. When wind conditions eventually allowed, he was winched aboard the CareFlight helicopter and transferred to Westmead Hospital. The activity leader's ankle was seriously fractured. He required orthopaedic surgery to insert plates and pins in the right ankle, and a closed reduction in the left. He is expected to use a wheelchair for 3 months or so. The previous day, the activity leader had informed their field manager that they were feeling physically exhausted from the program the previous week and asked whether they could take leave. However, they were informed that an alternative activity leader was not available to cover the activity, due to two activity leaders leaving the organization in the last month.

B.1 Coding Task Type 1: Checklist

Instructions to participants: Use the checklist below to select the categories from the UPLOADS classification scheme that best describe the contributory factors that were reported to play a role in this incident.

State and Federal Government
- ☐ Communication
- ☐ Funding and budgets
- ☐ Infrastructure and land management
- ☐ Policies and legislation
- ☐ Other

Regulatory Bodies and Associations
- ☐ Accreditation/licensing
- ☐ Auditing
- ☐ Communication
- ☐ Curriculum of outdoor education/recreation qualifications
- ☐ Funding and budgets
- ☐ Interactions with government
- ☐ Standards and code of practice
- ☐ Other

Local Area Government
- ☐ Auditing
- ☐ Communication
- ☐ Funding and budgets
- ☐ Legal responsibility for safety within the council area
- ☐ Policies and procedures
- ☐ Other

Schools
- ☐ Communication
- ☐ Dropping off/picking up participants
- ☐ Judgement and decision making
- ☐ Legal responsibility for safety of staff and students
- ☐ Planning and preparation for activity/trip
- ☐ Policies and procedures
- ☐ Teacher/student ratio
- ☐ Other

Parents/Carers
- ☐ Communication
- ☐ Dropping off/picking up participants
- ☐ Judgement and decision making
- ☐ Legal responsibility for safety of child
- ☐ Planning and preparation for activity/trip
- ☐ Other

Higher-Level Management
- ☐ Communication
- ☐ Financial constraints
- ☐ Judgement and decision making
- ☐ Organizational culture
- ☐ Policies and procedures for activities and emergencies
- ☐ Risk assessment and management
- ☐ Staffing and recruitment
- ☐ Supervision of staff (e.g. activity leaders, field managers)
- ☐ Supervision/oversight of activities and programs
- ☐ Training and evaluation of staff (e.g. activity leaders, field managers)
- ☐ Other

Supervisors/Field Manager
- ☐ Activity or program design
- ☐ Communication
- ☐ Compliance with procedures, violations & unsafe acts
- ☐ Experience, qualifications, competence
- ☐ Judgement and decision making
- ☐ Mental and physical condition
- ☐ Planning & preparation for activity
- ☐ Supervision of activity leaders and other staff
- ☐ Supervision/oversight of programs/activities
- ☐ Other

Activity Leader
- ☐ Communication, instruction & demonstration
- ☐ Compliance with procedures, violations & unsafe acts
- ☐ Experience, qualifications, competence
- ☐ Judgement and decision making
- ☐ Mental and physical condition
- ☐ Planning & preparation for activity/trip
- ☐ Situation awareness
- ☐ Supervision/leadership of activity
- ☐ Other

Activity Participant
- ☐ Communication & following instructions
- ☐ Compliance with procedures, violations & unsafe acts
- ☐ Experience & competence
- ☐ Judgement and decision making
- ☐ Mental and physical condition
- ☐ Planning & preparation for activity/trip
- ☐ Situation awareness
- ☐ Other

Other People in Activity Group
- ☐ Communication & following instructions
- ☐ Compliance with procedures, violations & unsafe acts
- ☐ Experience, qualifications, competence
- ☐ Judgement and decision making
- ☐ Mental and physical condition
- ☐ Planning & preparation for activity/trip
- ☐ Situation awareness
- ☐ Supervision of activity
- ☐ Other

Activity Group Factors
- ☐ Communication within group
- ☐ Group composition
- ☐ Group dynamics
- ☐ Group size
- ☐ Late arrival of group
- ☐ Teamwork
- ☐ Time pressure
- ☐ Other

Other People in Activity Environment
- ☐ Communication
- ☐ Compliance with procedures, violations & unsafe acts
- ☐ Experience, qualifications, competence
- ☐ Judgement and decision making
- ☐ Mental and physical condition
- ☐ Planning & preparations
- ☐ Situation awareness
- ☐ Other

Activity Equipment and Resources
- ☐ Documentation
- ☐ Equipment, clothing and Personal Protective Equipment
- ☐ Food & drink
- ☐ Medication (for those involved in the activity)
- ☐ Other

Activity Environment
- ☐ Animal & insect hazards
- ☐ Infrastructure & terrain
- ☐ Trees and vegetation
- ☐ Water conditions
- ☐ Weather conditions
- ☐ Other

B.2 Coding Task Type 2: Identify and Classify Contributory Factors

Instructions to participants: Using the table below, write a list of the contributory factors that were reported to play a role in this incident. For each factor that contributed to the incident, write the category from the UPLOADS classification scheme that best describes that factor.

Factors Contributing to the Incident	Category from the Classification Scheme
Factor 1	Category 1
Factor 2	Category 2
Factor 3	Category 3
Factor 4	Category 4
Factor 5	Category 5
Factor 6, etc.	Category 6, etc.

B.3 Coding Task Type 3: Classify Contributory Factors

Instructions to participants: The contributory factors that were described in the report are presented in the table below. For each factor that contributed to the incident, write the category from the UPLOADS classification scheme that best describes that factor.

Factors Contributing to the Incident	Category from the Classification Scheme
Activity leader feeling tired	Category 1
Hot day	Category 2
Activity leader forgot to wear helmet	Category 3
Activity leader violated company procedures by not wearing a helmet	Category 4
Activity participant failed to warn activity leader that he was not wearing his helmet	Category 5
Activity leader incorrect use of quickdraw	Category 6
Activity leader incorrect use of quickdraw	Category 7
Activity participant lacked experience as he did not identify the leader's lack of helmet or incorrect clipping	Category 8
Activity leader became anxious during the climb	Category 9
The program was poorly designed: a long day of multiple climbs with only a short break for lunch	Category 10
Field manager failed to act upon activity leader's concerns about physical exhaustion	Category 11
Not enough staff were available to cover the activity leader taking a break	Category 12
New staff had not been recruited to replace staff who had left	Category 13

Appendix C: UPLOADS Incident Report Form

 UPLOADS Incident Report

Section 1: INCIDENT CHARACTERISTICS

Incident reporter:			
Reporter present during incident? ☐ No ☐ Yes	Date of incident	Time of incident (24hrs)	State/Territory
Staff responsible for supervision at the time of the incident		Type of incident ☐ Near miss ☐ Adverse Outcome	
Actual Severi ty rating (0-6, see scale)		Potential Severity rating (0-4, see scale)	
Activity associated with incident		Main goals associated with activity	
Weather at the time of the incident Rain Conditions: Fine 1 2 3 4 Wet Temperature: Hot 1 2 3 4 Cold Wind conditions: Calm 1 2 3 4 Windy		Number of people involved in activity _____Participants (e.g. students) _____Activity leaders (e.g. instructors, guides) _____Supervisors (e.g. teachers) _____Volunteers (e.g. parents)	
Location of incident		Did the activity leader/s have relevant activity qualifications? ☐ No ☐ Yes	

Section 2: ADVERSE OUTCOMES (Not applicable for near misses)

2.1 Details of person impacted (if more than one person impacted, copy and paste this section)		
Name	Was the incident fatal? ☐ No ☐ Yes	
Experience in activity associated with incident ☐ Unknown ☐ No prior experience ☐ Some prior experience ☐ Extensive prior experience		
Injury type	Injury location	Illness
☐ Burns and corrosions ☐ Crushing injury ☐ Dislocation, sprain or strain ☐ Effects of foreign body entering through natural orifice ☐ Fracture ☐ Frostbite ☐ Injury to internal organs ☐ Injury to muscle, fascia or tendon ☐ Injury to nerves or spinal cord ☐ Open wound ☐ Poisoning by drugs, medicaments and biological substances ☐ Sequelae of injuries, of poisoning and of other consequences of external causes ☐ Superficial injury (e.g. abrasion, blister, insect bite) ☐ Toxic effects of substances chiefly nonmedicinal as to source ☐ Traumatic amputation ☐ Other and unspecified effects of external causes	☐ Head ☐ Neck ☐ Chest/Thorax ☐ Abdomen, lower back, lumbar spine and pelvis ☐ Shoulder and upper arm ☐ Elbow and forearm ☐ Wrist and hand ☐ Hip and thigh ☐ Knee and lower leg ☐ Ankle and foot ☐ Multiple body regions ☐ Unspecified part of trunk, limb or body region	☐ Abdominal problem ☐ Allergic reaction ☐ Altitude sickness ☐ Asthma ☐ Chest pain ☐ Diarrhoea ☐ Eye infection ☐ Food poisoning ☐ Hypothermia ☐ Heat stroke ☐ Menstrual ☐ Non-specific fever ☐ Skin infection ☐ Respiratory ☐ Urinary tract infection ☐ Unknown ☐ Other
Briefly describe any social/psychological impacts	Briefly describe any treatment at the scene of the incident	

Evacuation method	Hospitalisation required?	Emergency services called?
☐ Boat ☐ Helicopter ☐ Ski patrol-stretches ☐ Sled ☐ Stretcher ☐ Snowmobile ☐ Vehicle ☐ Walked out ☐ Not required	☐ No ☐ Yes	☐ No ☐ Yes Specify:

2.2 Overdue or missing people

Names	Emergency services called? ☐ No ☐ Yes, Specify:

2.3 Equipment loss/damage

Description of damage

2.4 Environmental damage

Description of damage

Section 3: INCIDENT DESCRIPTION

Describe the incident in detail. Include **who** was involved, **what** happened, **when** it happened and **where** it happened and any **equipment** involved. Do **not** enter identifying information (e.g. names)

Describe any relevant events leading up to the incident

Describe why the incident was a near miss (e.g. the activity leader pushed the participant out of the way just in time)

Section 4: CAUSAL FACTORS AND RELATIONSHIPS

Reporter: Explain in detail what you think caused the incident, including any relationships between the causes identified. Include any suggestions, comments or recommendations.

Manager: Explain in detail what you think caused the incident, including any relationships between the causes identified. Include any suggestions, comments or recommendations.

Definitions

An **'adverse outcome'** is defined as an event resulting in a negative impact, including: missing/overdue people; equipment or environmental damage; injury; illness; fatality; or social or psychological impacts.

A **'near miss'** is defined as a serious error or mishap that has the potential to cause an adverse event but fails to do so because of chance or because it is intercepted. For example, during a rock climbing activity an instructor notices that a participant's carabineer was not locked. If the student had fallen, this may have led to a serious injury.

Incident Severity Scale

	Severity Rating	Definition for Actual Severity Ratings*	Definition for Potential Severity Ratings**
0	No impact	Requires no treatment.	An incident where the potential outcome has a negligible consequence.
1	Minor	Requires localized care (non-evac) with short-term effects.	An incident where the potential outcome to risks has a low consequence.
2	Moderate	Requires ongoing care (localized or external, i.e. evac or not) with short- to medium-term effects.	An incident where the potential outcome to risks can cause moderate injuries or illnesses.
3	Serious	Requires timely external care (evacuation) with medium- to long-term effects.	An incident where the potential outcome to risks encountered is such that it may cause major irreversible damage or threaten life.
4	Severe	Requires urgent emergency assistance with long-term effects.	An incident where the potential outcome to risks encountered is certain death.
5	Critical	Requires urgent emergency assistance with serious ongoing long-term effects.	NA
6	Unsurvivable	Fatality.	NA

*Rate the Actual Severity of the incident in terms of the actual outcome of the event.
**Rate the Potential Severity of the incident in terms of the worst possible outcome, given the scenario.

Examples of Casual Factors

It is very important that you identify all the factors, and the relationships between them, which may have contributed to the incident you are reporting. To assist you in thinking about the causal factors involved in your incident, we have provided examples below of factors that have been found to play a role in previous incidents.

Activity Equipment and Resources
Documentation
Equipment, clothing, and Personal Protective
  Equipment
Food & drink
Medication

Activity Environment
Animal & insect hazards
Infrastructure & terrain
Trees and vegetation
Water/Weather conditions

Activity Leader/ Activity Participants/ Other People in Activity Group (e.g. teachers, parents, volunteers)
Communication, instruction & demonstration
Compliance with procedures, violations &
  unsafe acts
Experience, qualifications, competence
Judgement and decision making
Mental and physical condition
Planning & preparation
Situation awareness
Supervision/leadership of activity

Group Factors
Communication within group
Group composition
Group dynamics
Group size
Late arrival of group
Teamwork
Time pressure

Other People in Activity Environment (e.g. members of the public, emergency services)
Communication
Compliance with procedures, violations, &
  unsafe acts
Experience, qualifications, competence
Judgement and decision making
Mental and physical condition
Planning & preparation
Situation awareness

Supervisor/Field Managers
Activity or Program design
Communication
Compliance with procedures, violations, &
  unsafe acts
Experience, qualifications, competence
Judgement and decision making
Mental and physical condition
Planning & preparation for activity
Supervision of activity leaders and other staff
Supervision/oversight of programs/activities

Higher-Level Management
Communication
Financial constraints
Judgement and decision making
Organizational culture
Policies and procedures for activities and
  emergencies
Risk assessment and management
Staffing and recruitment
Supervision of staff
Supervision/oversight of activities and programs
Training and evaluation of staff

Schools
Communication
Dropping off/picking up participants
Judgement and decision making
Legal responsibility for safety of staff and
  students
Planning and preparation for activity/trip
Policies and procedures
Teacher/student ratio

Local Area Government
Auditing
Communication
Funding and budgets
Legal responsibility for safety within the
 council area
Policies and procedures

**Regulatory Bodies and Professional
 Association**
Accreditation/licensing
Auditing
Communication
Curriculum of outdoor education/recreation
 qualifications
Funding and budgets
Interactions with government
Standards and code of practice

Parents/Carers
Communication
Dropping off/picking up participants
Judgement and decision making
Legal responsibility for safety of child
Planning and preparation for activity/trip

State and Federal Government
Communication
Funding and budgets
Infrastructure and land management
Policies and legislation

Appendix D: Training Manual: The UPLOADS Approach to Accident Analysis

The following is a manual that was provided to organizations using UPLOADS to support their understanding of the systems thinking approach, and the design of the incident reporting system. It was supported by online training videos and face-to-face workshops.

D.1 The Theory behind UPLOADS

What will I learn in this manual?

- The principles of the systems approach;
- How the systems approach differs to other approaches;
- Rasmussen's (1997) risk management framework; and
- How Rasmussen's (1997) risk management framework can be used to analyze incident data from the led outdoor activity domain.

Why is this information important?

- The systems approach, and Rasmussen's (1997) risk management framework, underpin all aspects of the UPLOADS project from the development of the contributory factor classification scheme, and database, to the analysis of the industry dataset.
- We want the systems approach to underpin your approach to accident analysis and prevention.
- We want **you** to use this approach when you collect data, analyze the incident reports, and develop incident prevention strategies.
- Accident analysis not underpinned by an appropriate method and theory can do more harm than good.
- The systems approach will enable a more holistic understanding of what is causing incidents, and will inform the development of more appropriate and far reaching, incident prevention strategies.

D.1.1 The Principles of the Systems Approach

The systems approach involves three core principles.

Firstly, **behavior and safety is impacted by the decisions and actions of everyone in the system, not just frontline workers alone.** In the outdoor activity context, this means that decisions and actions made by politicians, CEOs, managers, safety officers, and work planners play a role in accidents, just as those made by instructors and participants do. This also means that safety is the shared responsibility of everybody working within the led outdoor activity system.

Second, **near misses and adverse events are caused by multiple, interacting, contributing factors, not just a single bad decision or action.** For example, a flawed decision made by an instructor that led to an accident will likely have various upstream contributory factors related to things like participants, training, procedures, management, equipment, program planning, etc. This means that there is no root cause of an incident, and that human error should never be seen as the cause of an incident. Rather, we need to search for the reasons as to why that error occurred. It also means that the relationships between contributory factors are as important to take into account as the factors themselves.

Third, **effective incident prevention strategies focus on systemic changes rather than individuals.** This means that strategies should generally focus on policies, procedures, and infrastructure rather than on punishment, warnings, or retraining. While changes to training programs at times may be appropriate, we need to recognize that it is very difficult to change individual behavior, especially if the system does not support changes in behavior. It is also not enough just to change the procedures and expect behavior to change. We need to examine the factors that may potentially impact on the execution of those procedures – such as staffing, management, or equipment availability.

Finally, as it is underpinned by the systems approach, the goal of UPLOADS is not to assign blame. Rather, we want to identify how factors across the led outdoor activity system combine to create injury causing incidents. In order to encourage people to report incidents, you need to keep this in mind at all times – the goal of UPLOADS is to learn from incidents, never to assign blame to individuals.

D.1.2 How the Systems Approach Differs from Other Approaches

The systems approach differs from other approaches in several ways.

The key points are, firstly, that **human error is seen as the outcome of an incident, rather than the cause of incidents.** This is in contrast to person-based approaches which focus on who made the error rather than why.

Second, the **systems approach is interested in how the interactions between parts of the system lead to errors**. For example, how an inadequate training program and limited availability of equipment might interact to shape an instructors performance in a way that leads to an incident. This is in contrast to approaches that focus on the 'hunt for the broken component' – such as root cause analysis. There the focus is on the instructor and their flawed performance, not on the factors that interacted to create the flawed performance.

Overall, the person approaches and other approaches which focus on component failures discourage reporting because they focus on blame, and fail to address the underlying, system-wide causes of incidents.

D.1.3 Rasmussen's (1997) Risk Management Framework

Accident causation models are a way of representing beliefs about how incidents occur. A model helps you determine what causes to look for and brings order to the way that you investigate accidents.

Rasmussen's (1997) risk management framework, shown in Figure D.1, was selected for the development of the UPLOADS project for a number of reasons. Firstly, it is domain-generic, so it can be readily applied to many different contexts, including outdoor activity provision. Second, it considers the entire led outdoor activity 'system', from the government to the activity environment.

Rasmussen's framework is underpinned by the idea that systems comprise various levels; actions and decisions across these levels interact with one another to shape behavior, safety, and accidents. Typically, the following system levels are described:

- The *government* level at which laws and regulations are developed;
- The *regulatory* level at which industry standards are developed based on laws and regulations;
- The *company* level where company policies and procedures based on industry standards govern work processes;
- The *management* level where company policies and procedures are implemented;
- The *staff* level representing the activities and characteristics of workers performing the processes; and
- The *work* level representing the equipment and environment within the work context.

In terms of accident causation, the framework argues that decisions and actions at all levels of the system interact with one another to shape system performance: safety and incidents are thus shaped by the decisions of all actors, not just the front line workers in isolation, and incidents are caused by multiple contributing factors, not just one bad decision or action.

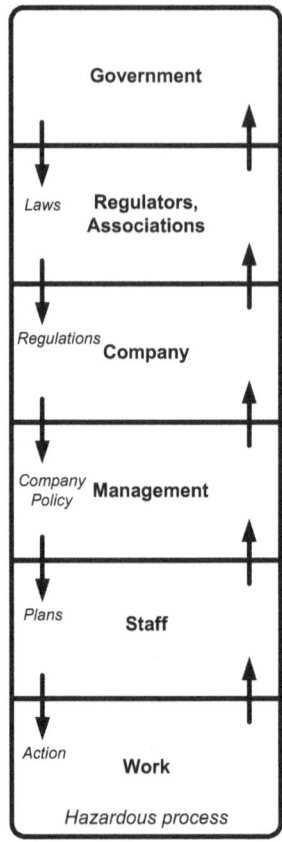

FIGURE D.1
Rasmussen's risk management framework. (Adapted from Rasmussen, J. 1997. *Safety Science,* 27(2/3), 183–213. With permission.)

The model also argues that for safe and efficient performance, the decisions and actions made at higher governmental, regulatory, and managerial levels of the system should propagate down and be reflected in the decisions and actions occurring at the lower levels. Conversely, information at the lower levels regarding the system's status needs to transfer up the hierarchy to inform the decisions and actions occurring at the higher levels. This is known as 'vertical integration' and is a key component of safe system performance.

We have adapted Rasmussen's framework to describe the led outdoor activity 'system' as follows:

- *Government policy and budgeting.* This refers to the government activities, decisions, actions, etc., relating to the provision of led outdoor activities.

- *Regulatory bodies and associations.* This level refers to the activities, decisions, actions, etc., made by personnel working for led outdoor activity regulatory bodies or associations.
- *Local area government; Activity centre management; Planning and budgeting; schools and parents.* This level refers to the activities, decisions, actions, etc., made by personnel working in local government, at the senior managerial levels of the activity centre involved (e.g. executive board level), at the schools involved, and by parents of the participants involved in the incident. These factors are related to higher level management, planning, and budgeting activities and typically occur before the incident itself (this can even be years preceding the incident).
- *Supervisory and management decisions and actions.* This level refers to the activities, decisions, actions, etc., made by personnel at the supervisory and managerial levels of the organization providing the activity involved in the incident. These factors typically occur prior to the incident itself but can also include decisions and actions made during, or in response to, the incident.
- *Decisions and actions of leaders, participants and other actors at the scene of the incident.* This level refers generally to the activities undertaken 'at the sharp end' prior to, and during, the incident. It therefore describes the flow of events leading up to and during the incident in question. This includes decisions and actions made by instructors and participants, but it may also include decisions and actions made by other actors, such as supervisors, emergency responders, members of the public, etc.
- *Equipment and surroundings.* The equipment and surroundings level refers to factors associated with the equipment used in support of the activity, the physical environment in which the activity was undertaken, and the ambient and meteorological conditions prior to or during the incident.

In the UPLOADS project, this framework underpins the classifications scheme that is provided to categorize the causal factors involved in incidents.

In conjunction with this framework, Rasmussen developed the Accimap technique to graphically representing the conditions that produce accidents (Rasmussen, 1997; Svedung & Rasmussen, 2002). Using Accimap involves constructing a causal diagram of the components, decisions, and actions that interacted with one another to create the system in which the accident in question occurred as well as the relationships between them.

As an example, the following section describes how this framework applies to the analysis of a major outdoor activity incident – the Mangatepopo Gorge Incident.

Example: Mangatepopo Gorge Incident Accimap Analysis

Here we present a description of the Mangatepopo incident to demonstrate how Rasmussen's risk management framework and associated Accimap method can be used to describe the causal mechanisms involved in led outdoor activity incidents (Salmon, Cornelissen, & Trotter, 2012).

The Mangatepopo tragedy occurred on the 15th April 2008 when a group of ten college students and their teacher, led by an instructor from an Outdoor Pursuit Centre (OPC), were completing a gorge walking activity in the Mangatepopo Gorge in the Tongariro National Park, New Zealand. Due to heavy rain in the area, a flash flood occurred which led to increased river flow and a rising river level in the gorge. As a result, the group had to abandon the gorge walking activity and became trapped on a small ledge above the water. Fearing the group would be washed off the ledge, the instructor decided to attempt to evacuate the group from the ledge and gorge by entering the river, with poor swimmers tied to stronger swimmers, following which the instructor would extract them downstream using a 'throwbag' river rescue technique (whereby a bag attached to a length of rope is thrown to the person in the water and used to pull them to the river-bank). After initiating the evacuation plan, only the instructor and two students managed to get out of the river as intended, with the remaining eight students and teacher being swept downstream and then over a spillway. Six students and their teacher eventually drowned, with only two of those swept over the spillway surviving.

In the aftermath of the incident, the coroner and an independent investigation initiated by the activity centre involved identified various failures on behalf of the instructor, her manager, the activity centre itself, the local weather service, and government legislation and regulation (Brookes, Smith, & Corkill, 2009; Davenport, 2010).

We used the contributory factors identified in the investigation reports and placed them across the different Accimap levels, as shown in Figure D.2.

The Accimap therefore shows how decisions, actions, and failures across the entire system interacted to enable the tragedy to occur. Importantly, it shows how factors from other levels of the system – including legislation and regulation, activity centre operation, instructor supervision, equipment, and environmental conditions – influenced the instructor's behavior and decision making during the incident. Thus, rather than lay the blame explicitly on the instructor, the description instead supports appropriate systems reform whereby inadequate conditions that shape instructor and participant behavior are identified and removed.

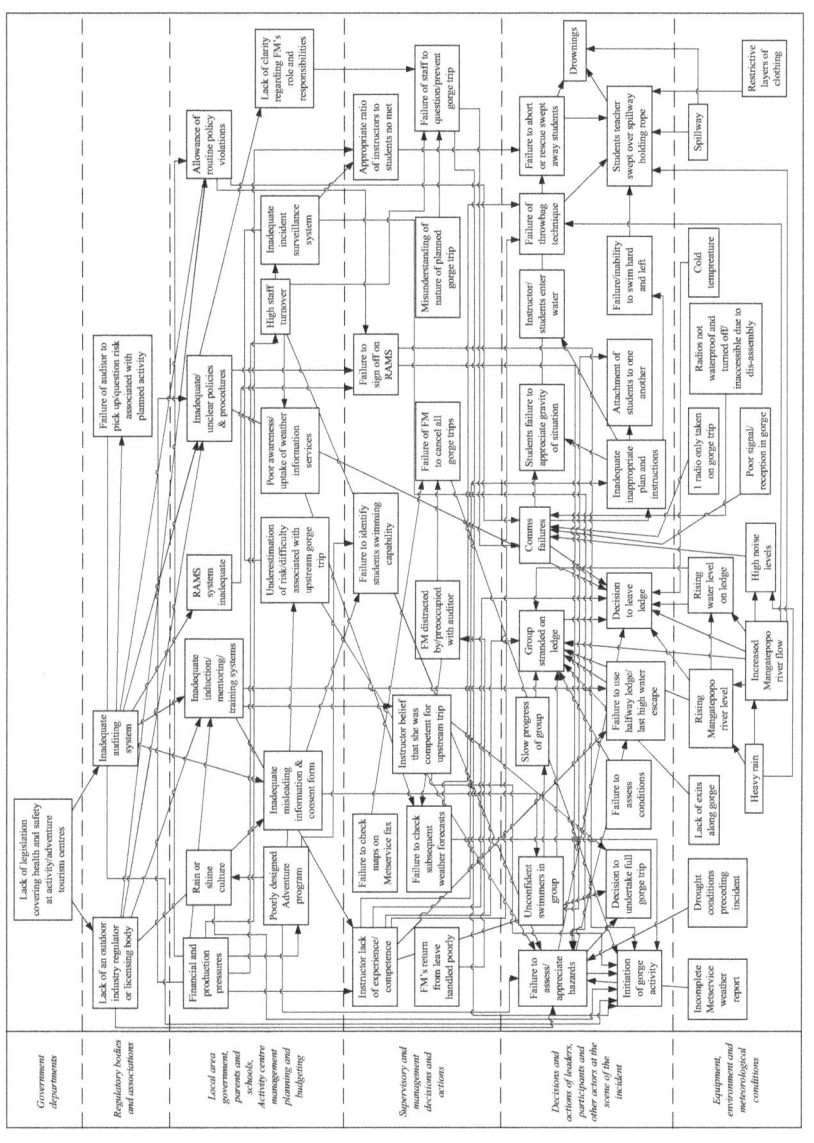

FIGURE D.2

Mangatepopo incident Accimap. (From Salmon, P. M., Cornelissen, M., & Trotter, M. J. 2012. *Safety Science*, 50(4), 1158–1170.)

Briefly, the Accimap depicts the failures across the led outdoor activity system that played a role. Starting at the bottom, examples of 'equipment and surroundings' factors include the adverse weather and conditions in the gorge, an incomplete weather report used in the morning staff meeting, and the radio used by the instructor (which was not waterproof and failed to work in the gorge due to poor reception).

Various 'physical processes and actor activities' were involved, including the instructor's decision to undertake the activity in the first place given the conditions, her flawed evacuation plan, and her failure to impart the gravity of the situation to the students and their teacher.

'Technical and operational management' failures shaped the instructor's performance on the day; she had limited experience of gorge-walking activities and lacked competence for them, both of which shaped her response to the unfolding incident. The centre's field manager failed to check the weather map on the reverse side of the faxed weather report (this showed the correct forecast), did not cancel all gorge trips in response to the adverse weather conditions, and failed to communicate his decision to cancel the downstream version of the trip.

'Company management' failures also played a role; Brookes et al. noted that, at the time, the centre was operating under financial and production pressures, which ostensibly contributed to a poorly designed adventure program, a rush to get staff trained and competent for activities, and the use of only one instructor for activities during busy periods. The centre's staff induction, mentoring and training programs, and risk-assessment and management systems, were also found to be inadequate. Other important factors identified included the centre's 'rain or shine' culture with regard to the conduct of activities in adverse weather conditions, high levels of staff turnover, and the lack of an effective accident and near-miss surveillance system.

At the 'regulatory factors' level, the absence of a regulatory or licensing body for outdoor activity centres at the time meant that unsafe practices could continue unchecked without reprisals.

Finally, at the 'government policy and budgeting level', a lack of legislation to oversee the provision of led outdoor activities also enabled the centre to continue engaging in unsafe practices.

All of the factors outlined in the Accimap combined in a way that enabled the tragic accident to happen. It is important to stress that the decisions and actions made on the day by those involved were shaped by various factors at the higher levels of the led outdoor activity system. Most of these were present long before the accident happened; without examination of existing practices and incident data, these failures remained unchecked and the activity centre continued to drift towards catastrophic failure.

D.2 Collecting Information about Incidents

This section presents some guidance on collecting information about incidents including:

- The paper-based incident report;
- The type of incidents to report;
- How to rate the severity of an incident; and
- How to write a good (useful) incident description

This information is included in the 'General Staff Member Training: How to Report an Incident' PowerPoint presentation.

To ensure that everyone knows how to report an incident, staff members should read the information contained in this section, or view the PowerPoint presentation.

D.2.1 Paper-Based Incident Report

In addition to the UPLOADS software tool, a paper-based form is provided to collect incident data from those involved in incidents (e.g. activity leaders, teachers, and participants).

This form should be made available to anyone who wants to report an incident. It may be useful to pin an example form on the noticeboard in your staff room, and email the form to all staff members.

From a practical point of view, it is better to encourage staff to fill in the form electronically and email it to you. This way you can copy and paste the details into the software tool.

D.2.2 What to Report

The UPLOADS software tool has the capability to record data on incidents involving:

- Adverse outcomes; and
- Near misses.

An **'adverse outcome'** is defined as an event resulting in a negative impact, including: missing/overdue people; equipment or environmental damage; injury; illness; fatality; or social or psychological impacts.

A **'near miss'** is defined as a serious error or mishap that has the potential to cause an adverse event but fails to do so because of chance or because it is intercepted. For example, during a rock climbing activity an instructor notices that a participant's carabineer was not locked. If the student had fallen, this may have led to a serious injury.

D.2.3 Incident Severity

A scale is provided to rate the severity of incidents.

Severity Rating	Definition for Actual Severity Ratings	Definition for Potential Severity Ratings
0 – No impact	Requires no treatment.	An incident where the potential outcome has a negligible consequence.
1 – Minor	Requires localized care (non-evacuation) with short-term effects.	An incident where the potential outcome to risks has a low consequence.
2 – Moderate	Requires ongoing care (localized or external, i.e. evacuation or not) with short- to medium-term effects.	An incident where the potential outcome to risks can cause moderate injuries or illnesses.
3 – Serious	Requires timely external care (evacuation) with medium- to long-term effects.	An incident where the potential outcome to risks encountered is such that it may cause major irreversible damage or threaten life.
4 – Severe	Requires urgent emergency assistance with long-term effects.	An incident where the potential outcome to risks encountered is certain death.
5 – Critical	Requires urgent emergency assistance with serious ongoing long-term effects.	NA
6 – Unsurvivable	Fatality.	NA

You need to rate each incident in terms of the *actual severity* and *potential severity*.

- Rate the ***actual severity*** of the incident in terms of the actual outcome of the event.
- Rate the ***potential severity*** of the incident in terms of the worst possible outcome, given the scenario.

So the data contained in the National Incident Dataset (NID) is not biased towards more serious events, it is important that you:

- Report any adverse outcome with an actual severity over 1; and
- Report any near miss with a potential severity over 2.

D.2.4 How to Write Good (Useful) Incident and Causal Factor Descriptions

What we can learn from UPLOADS is dependent on the quality of the data that we collect. The old adage applies: garbage in = garbage out. In order to fully understand the factors that contribute to incidents we need to gather as much detail as possible about the circumstances leading up to, during, and after the incident.

A good incident description will include a timeline of events that addresses the following questions.

Prior to the incident: Are there things that happened prior to the incident itself that you think influenced behavior in a way that enabled the incident to happen? For example:

- Did any events on the day contribute to the incident?
- What preparation or planning was undertaken to support the activity?
- Was this type of incident predicted in training or planning for the activity?
- Did other similar incidents occur prior to the one being reported?
- Were there flaws with the training programs, procedures, risk management systems, etc., used by your organization?
- Were activity programs sufficiently well designed?
- Any other details that you feel are relevant to the situation.

At the time of the incident:

- What activity was being undertaken?
- How many people were present (i.e. instructors, participants, teachers, volunteers, others), and who was participating in the activity?
- What was the weather like?
- What equipment was being used?
- Where adequate resources (equipment/staff) available to support the activity?
- Was adequate information available to support the activity (e.g. weather reports, maps, information on participant allergies, illnesses)?
- Were there any constraints that shaped how the activity proceeded (e.g. equipment and staff shortages)?
- Were appropriate communications taking place between activity centre staff?
- Any other details that you feel are relevant to the situation.

After the incident:

- What was the outcome of the incident or why was it considered a near miss?
- What treatment was provided at the scene?
- Was evacuation required? How did evacuation occur?
- Did treatment/evacuation run smoothly?
- Were adequate resources available for treatment/evacuation?
- Any other details that you feel are relevant to the situation.

You should then explain in detail what you think caused the incident, including any relationships between the causes identified. These conclusions should be based on the information you have provided in the incident description.

If possible, reporters and field managers should also make suggestions on how to prevent future, similar, incidents considering:

- What would have helped you understand the situation better?
- Would any specific training, experience, knowledge, procedures, or cooperation with others have helped?
- If a key feature of the situation was different, what would you have done differently?
- Could clearer guidance from your company have helped you make a better decision?

Remember: The information included in the incident description section of the report should not include any identifying information. Refer to the people involved in the incident by their role (e.g. participant, teacher, activity leader, camp organizer, or field manager).

The section on 'Incident Investigations' presents some general guidance on how to conduct more detailed incident investigations for serious adverse events or near misses. The information collected in these investigations can also be entered into the incident description and causal factor sections of the incident database.

D.2.5 Examples of Good and Bad Incident Reports

Example 1: Unopened Can Heated on Gas Stove

Bad:

Boys doing the wrong thing heating unopened can on camp stove. Tom received severe burns to face and cuts from shrapnel from exploding can.

Good:

It was the end of a long day of hiking. I was unpacking equipment from the car at the time of the incident. I had spoken to the two teachers present at the start of the day, and I was under the impression that they were going to supervise cooking the dinner. However, I am not sure now how clearly we discussed the issue, as they arrived at the campsite late, and we had little time to discuss our plans before we were scheduled to start the hike. Once we started hiking, we didn't get much time to discuss our plans as we were focused on making sure everyone was involved in the activity. There were 30 students and only 3 of us.

At the time of the incident, the participants were cooking dinner using camp stoves. The participant who was injured placed an unopened can of food directly over the gas. I am not sure whether he thought that was the correct way to heat the food or if there was some element of 'messing' around. It was a group of five boys, so potentially there was some element of showing off. The participant went to remove the can from the heat with a set of tongs – the can exploded, sending out piping hot food and shrapnel. Participant received burns and cuts to the face. No instructors were present at the time of the incident – we should have been supervising this activity, especially on the first day. There may have been some miscommunication – I was under the impression that the teachers were supervising dinner, but they were also absent at the time of the incident.

I gave the participant immediate first aid at the scene – running his face under cold flowing water. One of the teachers called an ambulance, and fortunately we only had to wait 30 minutes for it to arrive. One of the teachers went with the participant in the ambulance.

From my perspective the causal factors involved in the incident were: the participant doing the wrong thing (unsafe acts); the participant inexperience with heating up canned food on gas stove; showing off to the other boys; peer pressure – a group of boys egging each other on; miscommunication between instructors and teachers; I did not provide adequate instruction to the participants; the late arrival of school group at campsite; and a lack of allocation of responsibilities on the schedule.

Potentially, this problem could be avoided in the future by allocating responsibilities for supervision and organizing equipment on the schedule prior to arrival. That way teachers and instructors would both know what they were required to do.

Example 2: Unexpected Asthma Attack

Bad:

Participant had asthma attack while completing orienteering exercise.

Good:

The participants were completing an orienteering exercise in groups of three. Participant became short of breath while running and one of the group members came to get me (I was waiting at the end of the course). We ran back to where the participant was – she seemed to be having an asthma attack, but I wasn't sure and she couldn't speak. I asked her group members whether they were aware she had asthma – they said she had often had an asthma attack during running exercises at school. To my knowledge we had not been told by the school or parents that one of the participants had asthma. The girl didn't appear to have any medication with her, so I used the Ventolin inhaler out of the medical kit.

After about half an hour, she used the inhaler three times, her breathing calmed down, and we walked slowly back to camp. I gave her a glass of water and she sat quietly. She later told me that she had left her medication at home. I looked over the schedule, and we identified other activities that she thought might trigger another asthma attack.

I did notice earlier in the day that she had been reluctant about the orienteering exercise, but she didn't tell me why at the time. I later looked at her records, and there is no mention of any medication or asthma.

Potential causal factors: lack of information from school/parents, participant pre-existing medical condition, lack of medication, incorrect documentation, participant information not communicated.

This sort of incident has happened before – maybe we need to start asking participants if they have any medical conditions or injuries during the safety briefing, rather than relying on the consent forms from parents.

Example 3: Incorrect Use of Abseiling Equipment

Bad:
Participant didn't use the prussick properly to slow down while abseiling.
Good:
Participant was using the abseil rack for a third descent – he chose to adjust it to go faster. He went faster than he anticipated, and instead of letting go of the prussic, he held onto it tighter and therefore didn't slow down. He finally let go and the prussick stopped his descent, but by this time had a reasonable rope burn to his brake hand. He continued the activity for another two hours – no first aid was required.

Overall, the participant did not have the experience to know what to do when moving too fast. Next time I would place a safety line onto each person and retain control rather than rely on the participant to arrest themselves with a prussick back up.

D.3 Incident Investigations

For more serious incidents or near misses you may want to undertake an investigation to gather more information about what happened. This section presents some general guidance on how to conduct investigations using the systems thinking approach. The information collected in these investigations can also be entered into the narrative section of the incident report form.

For readers who would like to develop their investigation skills further we recommend: *The Field Guide to Understanding Human Error* by Sidney Dekker (2002). The advice presented in this section is drawn from Dekker's book, which presents practical guidance on how to conduct incident investigations in the workplace. Although the examples are drawn from healthcare, aviation, and industrial safety, the general principles also apply to the provision of outdoor activities.

D.3.1 Guiding Principles

Investigations need to be driven by one guiding principle: human errors are never an explanation for incidents (Dekker, 2002). Investigations need to go beyond what people should or could have done. Instead, we need to know 'why' they did what they did. Why did their actions or decisions *make sense at the time?*

This means that you are only interested errors, mistakes, or violations to the extent that they can tell you about the system. You need to ask the questions:

- What are the sources of people's difficulties?
- What are trade-offs were being made in that situation?
- What influence was technology or equipment having?
- What workload was present in the situation?
- How did communication help or hinder them? What uncertainties were there in the environment?

D.3.2 Potential Sources of Information about Incidents

Several sources of information can be used to gather information on why incidents happened:

- The documents that supported the activity (e.g. plans, schedules, checklists, standard operating procedures);
- Activity leader training manuals or guidelines;
- The documents that were used during the activity (e.g. equipment manuals and maps); and
- Interviews with the people directly involved in the incident and those that planned and supported the activity.

D.3.3 Interviewing People

The people involved in the event are a key source of information. Interview those involved in the planning of the activity as well as those directly involved at the time of the incident.

A suggested method for interviews:

1. Start by telling them that the purpose of the interview is not to assign blame, but to understand why the incident happened, and prevent similar future incidents from occurring.

2. Let them tell the story from their perspective, without interruption for clarification.

3. Tell the story back to them, and ask questions to clarify points you don't understand or where pieces of information may be missing. If they don't know 'why' they made a certain decision, that's fine. Move on.

4. Help them to construct a timeline of events. At critical points in the timeline probe:

   - Cues: What where you looking at? What were you expecting to happen?

   - Knowledge: What knowledge were you using to deal with the situation? Had you had any experience with similar situations?

   - Goals: What were you trying to achieve at the time? Were there conflicts between your goals? Were you under any time pressure?

   - Influences: How did other influences (in the environment or organization) help determine how they interpreted the situation and how they acted?

   - Outcome: Was the outcome expected? Did you have to revise your assessment of the situation?

These questions are but a guide – with practice you will find your own way of gathering this information and asking these questions in your own words. Generate your own questions to gather the information that you feel is appropriate to your organization and the activities you are running.

D.3.4 Dealing with Inconsistencies

It is quite likely that different people will remember the event differently. That's okay – there is no one 'true' course of events. Each perspective may have valuable pieces of puzzle to add to the 'why' behind the incident. Dekker (2002) suggests some strategies for dealing with inconsistencies:

- Make the disagreement and inconsistencies explicit;

- If later statements from the same people are contradictory, decide which version you want to rely on and state why; and

- Do not see disagreements and inconsistencies as 'problems'. Potentially, they are also contributing factors to the incident. For example, they may point to conflicts between the goals of different people in the situation.

D.3.5 Traps to Avoid

Understanding incidents from the perspective of those involved is a tricky business. It's important not to become focused on 'if only' (e.g. if only they had paid attention to the participant or decided not to go down the track, then the incident would never have happened). When this happens, focus shifts from *learning* to *blame*. Dekker (2002) highlights some common traps to avoid when collecting data:

- **Hindsight** biases your investigation towards issues that you *already know* are important. As a result, you may assess people's behavior in light of what they *should* have known. You need to try and understanding the evolving situation from the point of view of the people who didn't know the outcome, to see why their actions and decisions made sense at the time.

- The **'root cause'** of an incident doesn't matter nor make sense. Typically, risk management plans include multiple defences against incidents occurring. Thus, multiple failures have to occur before these defences are broken. Expect to find multiple contributing or causal factors when investigating an incident.

- **Cherry picking** evidence to support initial hypotheses about the causes of incidents can lead to biased investigations. Generate your hypotheses, and then look for evidence that doesn't support your hypotheses to test them.

References

Brookes, A., Smith, M., & Corkill, B. (2009). *Report to the Trustees of the Sir Edmund Hillary Outdoor Pursuit Centre of New Zealand: Mangatepopo Gorge Incident, 15th April 2008.* Retrieved from http://www.hillaryoutdoors.co.nz/newsite/wp-con tent/uploads/2013/06/091015-IRT-OPC_-Report.pdf

Davenport, C. J. (2010). *Mangatepopo coroner's report.* Retrieved from http://outdoor council.asn.au/doc/Coroners_Report_OPC.pdf

Dekker, S. (2002). *The field guide to understanding 'Human Error'.* Aldershot, UK: Ashgate.

Rasmussen, J. (1997). Risk management in a dynamic society: A modelling problem. *Safety Science, 27*(2/3), 183–213.

Salmon, P. M., Cornelissen, M., & Trotter, M. J. (2012). Systems-based accident analysis methods: A comparison of Accimap, HFACS, and STAMP. *Safety Science, 50*(4), 1158–1170.

Svedung, I., & Rasmussen, J. (2002). Graphic representation of accident scenarios: Mapping system structure and the causation of accidents. *Safety Science, 40*(5), 397–417.

Index

Page numbers followed by f and t indicate figures and tables, respectively.